Springer Series in Materials Science

Volume 334

The Springer Series in Materials Science covers the complete spectrum of materials research and technology, including fundamental principles, physical properties, materials theory and design. Recognizing the increasing importance of materials science in future device technologies, the book titles in this series reflect the state-of-the-art in understanding and controlling the structure and properties of all important classes of materials.

Vladimir A. Bershtein · Pavel N. Yakushev

High-Temperature Polymer Nanocomposites Based on Heterocyclic Networks from Nitrile Monomers

Springer

Vladimir A. Bershtein
Ioffe Institute
Russian Academy of Sciences
St. Petersburg, Russia

Pavel N. Yakushev
Ioffe Institute
Russian Academy of Sciences
St. Petersburg, Russia

ISSN 0933-033X ISSN 2196-2812 (electronic)
Springer Series in Materials Science
ISBN 978-3-031-32942-5 ISBN 978-3-031-32943-2 (eBook)
https://doi.org/10.1007/978-3-031-32943-2

This Springer imprint is published by the registered company Springer Nature Switzerland AG
The registered company address is: Gewerbestrasse 11, 6330 Cham, Switzerland

Preface

With the development of technology, the attention and the demands increase to lightweight high-temperature dielectric materials for work in extreme conditions at temperatures up to 300–400 °C and higher (at short-term exposure), in many fields, such as aircraft and space fields, marine industry, microelectronics, as well as for work in the extremely high-temperature zones of the various constructions and devices. The main direction in this regard is the creation of nanocomposites based on heat-resistant matrices and inorganic nanoparticles that enhance the properties of the polymer matrix.

For many years, mainly epoxy resins have been used as matrices for heat-resistant composites, in particular in aerospace applications. In recent decades, however, they are gradually losing their leading role in this respect since epoxy composites can still be used for a long time at temperatures not exceeding 200 °C.

As it is well known, among organic materials, the heterocyclic structures, in particular, heterocyclic polymer networks are the most thermally stable. Last decades, actually, four newly developed classes of resin matrices for high-temperature nanocomposites, which form heterocyclic networks (thermosets) after curing, have come to the market; they can be applied in composite materials without serious losses at temperatures above 200 °C. These are bismaleimide (BMI)-, benzoxazine (BOA)-, cyanate ester resin (CER)-, and phthalonitrile (PhN)-based matrices.

This book considers recent advances achieved during the last decade in physical studying two classes of these high-temperature nanocomposites, based on CER or PhN, whose network matrices, basically with carbon–nitrogen heterocycles, are formed by means of the polymerization of nitrile-containing compounds. These heterocyclic network-based nanocomposites have emerged as the best classes of polymers for applications under extreme conditions and are claimed to be the leading positions owing to their high thermal resistance and unique combination of other properties. The nanocomposites with different organically modified, "functionalized" nanoparticles (especially 3-D polyhedral oligomeric silsesquioxane (POSS) or SiO_2 nanoparticles, or 2-D montmorillonite (MMT) silicate nanolayers, etc.) have been studied; the functionalized nanofillers chemically interacted with the matrix networks forming the hybrid structures.

Three books on CER-based materials have previously been published (Hamerton I. Chemistry and Technology of Cyanate Ester Resins. Glasgow: Blackie Academic, 1994; A. M. Fainleib (Ed.). Thermostable Polycyanurates. Synthesis, modification, structure, and properties. Nova Science Publishers, New York, 2010, and Crawford A. Novel Cyanate Ester Blends. University of Surrey, 2014). Recently, two books on PhN-based materials (Mehdi Derradji, Jun Wang, Wenbin Liu. Phthalonitrile Resins and Composites. Elsevier, 2018, and A. Dhanya, S. Chandran, D. Mathew and C. P. Reghunadhan Nair. High-Performance Phthalonitrile Resins. Challenges and Engineering Applications. De Gruyter, 2019) were also published. However, these books are devoted mostly to the chemical and technological problems and deal basically with the early works on the mentioned materials, and only to a small extent to nanocomposites.

At the same time, numerous papers devoted to the study of the structure and physical properties of nanocomposites based on heterocyclic matrices derived from nitrile monomers—CER or PhN—have been published during just the last decade. The main focus of these studies was the comparative analysis of nanostructures of composites, their relaxation properties and elastic characteristics in a wide range of temperatures, and the thermal stability of the nanocomposites as well as their interrelationships. This book is an overview of these recent works devoted to these most promising high-temperature nanocomposites.

It should be indicated that the authors' research findings from 15 articles published by the authors with coworkers during the last decade contributed greatly to this book's content. Of the 114 figures included into the book, about half are from the authors' own publications. Over the last decade, the comprehensive physical studies of nanostructures, matrix dynamics, and composites' properties, and their relationships in these nanocomposites have been performed by the authors using transmission electron microscopy (TEM), mid- and far-infrared spectroscopy (IRS), energy-dispersive X-ray spectroscopy (EDXS), dynamic mechanical analysis (DMA), differential scanning calorimetry (DSC), thermogravimetric analysis (TGA), laser-interferometric creep rate spectroscopy (CRS), and other techniques.

A number of unique experimental results and practically important effects have been demonstrated in these studies. Among them, the following should be mentioned.

1. In the case of the synthesis of composites with chemically embedded SiO_2 units introduced by the sol-gel technology, it was possible, for the first time, to create subnanometer-sized silica nodes in the matrix in the absence of clusterization. The authors not only introduced the term "subnanocomposites" into the literature but also showed the superiority of the properties of polymer subnanocomposites over those of the corresponding composites with SiO_2 nanoclusters.
2. The possibility of quasi-regular distribution of subnanoparticles or molecularly dispersed nanoparticles in nanovolumes of the amorphous heterocyclic matrices has been shown experimentally (by a combination of TEM and EDXS analysis).
3. It was found out that the maximum positive influence of covalently embedded inorganic nano- or subnanounits on the properties of polymeric heterocyclic

matrices is observed at their ultra-low content of only 0.02–0.1 wt.%, and the nature of this effect was revealed.

4. It was shown that this was due to an abnormally large "constrained dynamics" effect in the matrix (constrained interfacial dynamics), its determinant role in suppressing the dynamics of the whole matrix, and the positive impact of this phenomenon on composite properties.
5. A pronounced dynamic heterogeneity in the glass transition of the studied nanocomposites was shown.
6. Certain correlations between the nanostructure, matrix dynamics, and the properties of the investigated composites were traced.
7. We have shown the possibility (after heat treatment under certain conditions) to completely suppress the relaxation dynamics in nanocomposites, including the glass transition, and to achieve practical constancy of the elastic modulus in the temperature range from 20° to 600 °C.
8. A uniquely high for polymer composites $T_g = 570$ °C was achieved in the case of phthalonitrile nanocomposites after their post-curing.
9. It was shown that some phthalonitrile nanocomposites can be recommended for continuous use at temperatures up to 400 °C, and for short-term use for dozens of minutes (with the maintenance of material integrity) at temperatures up to 500 °C in an air environment and up to 900 °C in the inert, oxygen-free, atmosphere.

Of course, each of the above four classes of high-temperature composite materials has its own advantages and disadvantages, but at the moment there is reason to believe that the phthalonitrile nanocomposites are the most obvious candidate for the leading position among the most thermostable polymer nanocomposites.

This book is intended for use by researchers of composite materials and specialists engaged in material selection for work in extreme conditions; for students specializing in materials science; for polymer physicists, and for university libraries.

In conclusion, we consider as a pleasant duty to express our appreciation and deep gratitude to our long-term coauthors, chemists from the Institute of Macromolecular Chemistry of the National Academy of Sciences of Ukraine, Corresponding Member of the Academy of Sciences of Ukraine Prof. A. M. Fainleib and his coworkers, O. P. Grigoryeva, O. M. Starostenko, K. G. Gusakova, and O. G. Melnychuk. These scientists synthesized and submitted all the composites for our physical studies; the obtained results were the content of our coauthored articles published during the period from 2012 to 2021.

In addition, we express deep gratitude to our colleagues and coauthors at the Ioffe Institute who have contributed significantly to these studies: D. A. Kirilenko, who has performed structural studies of nanocomposites by TEM and EDXS methods, and V. A. Ryzhov, who has carried out the experiments with far-IR spectroscopy.

St. Petersburg, Russia

Prof. Vladimir A. Bershtein
Dr. Pavel N. Yakushev

Contents

Abbreviations

1-D	One-dimensional
2-D	Two-dimensional
3-D	Three-dimensional
6F-DCBA	Dicyanate ester of hexafluorobisphenol A
AEAPIB-POSS	Aminoethylaminopropylisobutyl—POSS nanoparticles
Amino-MMT	Modified MMT nanolayers containing amino groups
APIB-POSS	Aminopropylisobutyl—POSS nanoparticles
Apph	4-Aminophenoxy phthalonitrile, curing agent
APTES	aminopropyltriethoxysilane
APTMS	aminopropyltrimethoxysilane
BACE1	At the synthesis of CER-Silica, CE : APTES : TEOS = 1: 0.5 : 1
BACE2	At the synthesis of CER-Silica, CE : APTES : TEOS = 1: 1 : 1
BACE3	At the synthesis of CER-Silica, CE : APTES : TEOS = 1: 1.5 : 1
BACE4	At the synthesis of CER-Silica, CE : APTES : TEOS = 1 : 2 : 1
Baph	2,2-bis [4-(3,4-dicyanophenoxy)phenyl] propane (monomer)
BAPhN	Bisphenol A-based phthalonitrile
BCC	1,3-bis(4-cyanatobenzyl) cyclohexane monomer
BCE	Neat matrix formed from 1,4-bis(2-(4-cyanatophenyl)-2-propyl)benzene monomer
BMI	Bismaleimide
BN	Boron nitride
BNNS	Boron nitride nanosheets
BOA	Benzoxazine
BPh	4,4'-Bis (3,4-dicyanophenoxy) biphenyl, monomer
CE, CY	Cyanate ester
CER	Cyanate ester resin
CER-A	CER network based on dicyanate ester of bisphenol-A
CER-E	CER network based on dicyanate ester of bisphenol-E
CNTs	Carbon nanotubes
CRS	Laser-interferometric creep rate spectroscopy
DCBA	Dicyanate ester of bisphenol-A

DCBE	Dicyanate ester of bisphenol-E
DMA	Dynamic mechanical analysis
DSC	Differential scanning calorimetry
DTG	Derivative thermogravimetry
ECH-POSS	Epoxy cyclohexyl-POSS nanoparticles
EDXS	Energy-dispersive X-ray spectroscopy
EIS	Electrochemical impedance spectroscopy
E-POSS	Epoxycyclohexyl-POSS nanoparticles
ETP	Eu-containing complexes for modifying carbon nanotubes
Far-IR	Far-infrared spectroscopy
FMCM-41	Glycidyl silane-functionalized mesoporous silica particles MCM-41
FTIR	Fourier transform infrared spectroscopy
GO	Graphene oxide
GONSs	Graphene oxide nanosheets
GPTMS	3-glycidoxypropyltrimethoxysilane
GX-540	Aminopropyltrimethoxysilane
HAADF-STEM	High-angle annular dark-field imaging in a scanning transmission electron microscope
h-BN	Hexagonal lattice boron nitride
HDPE	High-density polyethylene
IRS	Infrared spectroscopy
MAX phase	Ti_3AlC_2 or Ti_3SiC_2 ceramics nanoparticles
MCM-41	Mesoporous silica particles
Mid-IR	Mid-infrared spectroscopy
MMT	Montmorillonite
MWCNTs	Multiwalled carbon nanotubes
MXene phase	$Ti_3C_2(OH)_2$ ceramic nanosheets
NMR	Nuclear magnetic resonance
N-POSS	N-phenylaminopropyl—POSS nanoparticles
P(Baph)	Polymerized bisphenol-A phthalonitrile resin
PD	Polymerization degree
PhN	Phthalonitrile
POSS	Polyhedral oligomeric silsesquioxane
PP	Polypropylene
SANS	Small-angle neutron scattering
SAXS	Small-angle X-ray scattering
SEM	Scanning electron microscopy
SiN, α-Si_3N_4	Silicon nitride
SWCNTs	Single-walled carbon nanotubes
TEM	Transmission electron microscopy
TEOS	Tetraethoxysilane
T_g	Glass transition temperature
TGA	Thermogravimetric analysis
uMWCNTs	"Unzipped" multiwalled carbon nanotubes

W	Tungsten
WAXD	Wide-angle X-ray diffraction
xGnP	Exfoliated graphite nanoplatelets
XPS	X-ray photoelectron spectroscopy
XRD	X-ray diffraction

Chapter 1
Introduction. About Heat-Resistant Polymer Thermosets Used as Matrices for Nanocomposites

Abstract This introductory chapter gives a brief overview of the four classes of the most heat-resistant polymer matrices of nanocomposites capable of long-term service at temperatures of 250–300 °C and short-term service at 400 °C. Their approximate characteristics are given, including the attainable limit values of Tg, thermal stability, elastic modulus, dielectric constant, thermal conductivity, and other properties of nanocomposites based on them.

The development of new technologies in a number of fields necessitates the use of lightweight, durable, and high-temperature-resistant dielectric materials for structural elements and various hot-section components. This is of particular interest for microelectronics, last-generation aviation and space programs, for marine engineering, etc.

Until recent decades, epoxy resins were widely used for these purposes. In favor of their use were, in particular, their commercial availability and relatively low cost. Although epoxy resins with improved performance properties are being developed, they are currently losing their leading role as high-performance materials. This is due to their relatively low hot/wet performance: the possibility of long-term operation at temperatures not exceeding 200 °C and water absorption up to 5 wt.%. At the same time, at present in some cases, there is a need for long-term resistance of materials in some structures to temperatures of 300–400 °C and even higher, at short-term action.

High-performance polymers are a group of polymer materials that are known to retain their desirable mechanical, thermal, and other important properties when exposed to strong environmental influences such as high temperature, high pressure, high humidity, and irradiation.

By now, the point of view has been finally formed that four classes of high-heat-resistant thermosets-matrices for high-performance nanocomposites, developed mainly in the last two–three decades, are suitable for these purposes. These are thermosets based on such monomers as bismaleimide (BMI), benzoxazine (BOA), cyanate ester (CE), and phthalonitrile (PhN).

V. A. Bershtein and P. N. Yakushev, *High-Temperature Polymer Nanocomposites Based on Heterocyclic Networks from Nitrile Monomers*, Springer Series in Materials Science 334, https://doi.org/10.1007/978-3-031-32943-2_1

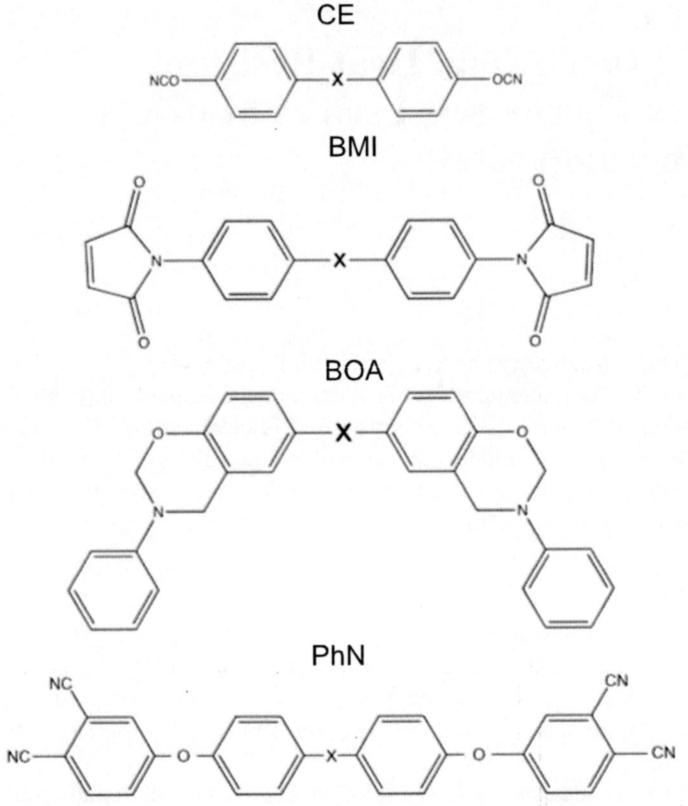

Fig. 1.1 Semi-schematic images of monomers for heterocyclic thermosets-matrices of high-temperature nanocomposites: Cyanate Ester (CE), Bismaleimide (BMI), Benzoxazine (BOA), and Phthalonitrile (PhN). The X sign corresponds to the different atomic groups, e.g., alkyl, aryl, etc.

Figure 1.1 shows semi-schematic images of these monomers for heterocyclic matrices of high-temperature nanocomposites. The polymerization of these thermosetting monomers results in the formation of densely cross-linked polymer networks that include various types of heterocycles, mainly carbon–nitrogen and also carbon–oxygen or carbon–nitrogen-oxygen ones. Thus, the polymerization of PhN leads primarily to the formation of a network with phthalocyanine heterocycles as well as isoindoline and triazine heterocycles in the forming network. Only triazine heterocycles are formed during CE polymerization. Schemes of the cycles are given in this book.

Thermosetting resins, which, when cured, form heterocyclic polymer networks can be applied as the matrices in composite materials without serious losses at high loads at temperatures above 200 °C. The first three of these matrices can be used at operating temperatures in the 250–300 °C range, while the PhN-based thermosets, as it will be shown later, exhibit unusually high thermal and thermo-oxidative stability:

after a high-temperature procedure of post-curing treatment, they can operate at temperatures up to 400 °C, exhibit T_g > 400 °C and a low mass loss temperature of $T_{5\%}$ > 500 °C. As it will be clear from the following book content, phthalonitrile nanocomposites with the highest thermal properties undoubtedly occupy a leading position in this respect.

The most detailed information about these four classes of highly heat-resistant polymer matrices can be found in books and reviews (Bershtein et al. 2010; Crawford 2014; Derradji et al. 2018; Dhanya et al. 2019; Fainleib 2010; Hamerton 1994; Iredale et al. 2017; Ishida and Agag 2011; Rimdusit et al. 2013; Stenzenberger 1990). Each group of these thermosetting matrices under consideration has, of course, its own advantages and disadvantages. The latter may be due to various reasons: poor processability, high post-curing temperatures, harmful initial raw materials, relatively high dielectric constant, a rather high cost, and low elongation at break, etc. As for the last of these disadvantages, the following should be mentioned: to diminish its impact, Babkin et al. (2015) obtained PhN monomers with flexible siloxane bridges, and their thermosets demonstrated high thermal and thermo-oxidative stabilities at the same level as described for that type of polymer matrices.

Among the main advantages of these high-performance matrices are, in particular,

- easy production technology, low water uptake, near-zero volumetric shrinkage upon polymerization, suppressed dynamics (low tan delta at DMA), good sound and noise absorbance, chemical and flammability resistance, and low-cost raw materials *for BOA-based matrices*;
- unusually low capacitance properties, low dielectric constant, low water uptake (down to 0.7 wt.%), excellent adhesion to metals at T ~ 250 °C, and reaching T_g = 400 °C in the case of CE oligomer use *for CER-based matrices*;
- high modulus and near-zero shrinkage, combining the high-temperature properties of thermosetting polyimides with the facile processability of the standard epoxy resins *for BMI-based matrices*;
- the "magic T_g" > 400 °C after post-curing, high mass loss temperature $T_{5\%}$, low dielectric constant and dielectric losses, and ray shielding efficiency
- *for PhN-based matrices.*

It is worth mentioning here the outstanding contribution of T. Keller and his colleagues. (Naval Research Laboratory, NRL, USA) to the synthesis and creation of various phthalonitrile matrices for high-temperature-resistant nanocomposites (Keller and Griffith 1980; Keller and Price 1982; Keller 1992, 1993; Keller and Dominguez 2005; Laskoski et al. 2007, 2014, 2015; Sastri et al. 1996, 1997). One of the most important results in the course of PhN thermosets research in NRL was the absence of softening PhN thermoset at 500 °C under the condition of post-curing in an inert atmosphere at high temperature (see also Chaps. 6 and 7 about complete suppression of the glass transition and the constancy of the elastic modulus over the temperature range 20–600 °C under these conditions).

Table 1.1 contains very approximate characteristics of the above four matrices taken from various sources including also the experimental data described in this book below.

Table 1.1 Approximate values of some physical properties of four main neat high-temperature-resistant matrices for polymer nanocomposites based on the indicated monomers (the data were taken from various sources)

Property	BMI	CE	BOA	PhN
Density, g/cm^3	1.28–1.34	1.1–1.35	1.12–1.19	1.3–1.4
Tensile strength, MPa	40–80	56–120	44–64	40–110
Tensile modulus, GPa	3.6–4.8	2.7–3,4	3.8–4.5	2.0–5.6
Breaking deformation, %	1.6–3.0	2.0–3.8	1.3–2.9	1.5–2.5
Dielectric constant, 1 MHz	3.1–3.7	2.7–3.3	3.0–3.6	2.2–2.8
Curing shrinkage, %	~ 0	~ 3	~ 0	–
5% mass loss temperature $T_{5\%}$, °C	430–480	280–410	275–462	460–580
Glass transition temperature T_g, °C	280–340	250–400	180–350	300–446
Coefficient of thermal expansion (ppm/°C)	45–50	54–71	58–69	40–60

Last years the scientists and engineers working in the fields under consideration try to find ways to optimize the synthesis methods, structure, and properties of these high-performance materials. One of the directions is to obtain copolymers of these thermosets. Due to the high chemical activity, the functional groups of such monomers/oligomers easily enter into chemical interaction with the formation of densely cross-linked heterocyclic copolymers, providing an optimal combination of the necessary characteristics for the efficient operation of these materials.

Really, in some works, attempts were made to further improve their properties by combining two or even three types of these thermosets in one material by their chemical "cross-linking" into a single network or by obtaining their physical mixtures. Thus, various combinations based on bismaleimide and benzoxazine (Agag and Takeichi 1424; Chaisuwan and Ishida 2006; Jin et al. 2010; Kumar et al. 2007; Liu and Yu 1890; Takeichi et al. 2008; Wang et al. 2015a), bismaleimide and CER (triazine cycles) (Crawford et al. 2016; Guan et al. 2011; Hamerton 1996; Hamerton et al. 2001; Hu et al. 2010; Ren et al. 2016; Wu et al. 2016, 2013; Yang et al. 2018; Yuan et al. 2014; Zeng et al. 2013), benzoxazine and CER (Chozhan et al. 2019; Kimura et al. 1113, 2013; Kumar et al. 2009; Wang et al. 2417), and bismaleimide–CER-benzoxazine combinations (Wang et al. 2015b, c) have been synthesized and investigated.

In some cases, such a combination has led indeed to further improvements in material properties. For example, Takeichi et al. (2008) found that the T_gs of these BOA-BMI alloys were higher than those of each resin due to the formation of their co-cross-linking. Hamerton (1996) found that the interpenetrating network polymer structures comprising CE and BMI components with T_g ~ 270 °C and the onset of thermo-oxidative degradation range from ~ 390 °C manifested, under certain conditions, after their co-curing T_g > 400 °C and the onset of degradation at 425 °C in air medium.

However, the most significant influence on the properties of high-temperature matrices, of course, was achieved due to the introduction of functionalized nanoparticles into them. As it will be clear from the following, the above properties can be significantly improved in nanocomposites in a number of cases. Thus, mechanical strength may increase in this case by 2–3 times, and sometimes it can even be increased by almost an order of magnitude. The temperature of small mass loss upon heating $T_{5\%}$ for some of the nanocomposites under consideration increases from 460 to 550 °C. This book will present examples where the glass transition temperature T_g changes (from neat matrix thermoset to nanocomposite) becoming ~ 50–100 °C higher, for example, increasing for the phthalonitrile materials from 446° to 560–570 °C. The dielectric constant can be reduced from 3 to 2, and the elastic modulus sometimes increases 2–5 times after the introduction of nanoparticles, in particular from 1.5 GPa to 7 GPa. The possibility of increasing the thermal conductivity by 26 times in the nanocomposite was also found.

As stated above, the considered high-temperature-resistant matrices and nanocomposites based on them are of particular interest for next-generation aviation and space programs, marine engineering, and for microelectronics, although they can, of course, be used in other cases for different hot-section components of structures and devices. To date, the discussed nanocomposites are already used in world practice for the manufacture of large parts of structures, such as the upper and lower wing skins and engine nacelle skins, satellite antennae, or gearboxes located near aircraft engines, as well as for the manufacture of strakes, fins, nose radomes, and different small parts (McConnell 2009). It should be mentioned that some types of modern aircraft composites make up about 20–30% of the weight of the entire structure.

Finally, it should be emphasized that the considered groups of high-performance thermosets are characterized not only by enhanced thermomechanical properties. As shown by Gu et al. (2022) in a review on potential applications of phthalonitrile thermosets in electronics, their applications include ray shielding materials, electromagnetic wave-transparent materials, electromagnetic interference shielding materials, supercapacitors, and magnetoresistance materials. The gamma rays shielding efficiency of the neat phthalonitrile thermoset and phthalonitrile/tungsten nanocomposites is described in Sect. 9.5. Therefore, the thermosets under consideration hold also great promise for performing in applications such as composites, coatings, adhesives, and encapsulants for use in the electronics industry exhibiting versatility in a wide range of applications.

References

T. Agag, T. Takeichi, Preparation, characterization, and polymerization of maleimidobenzoxazine monomers as a novel class of thermosetting resins. J. Polym. Sci. Part A: Polym Chem. **44**, 1424 (2006). https://doi.org/10.1002/pola.21245

A.V. Babkin, B.A. Bulgakov, A.V. Kepman et al., Low-melting siloxane-bridged phthalonitriles for heat-resistant matrices. Europ. Polymer J. **66**, 452 (2015). https://doi.org/10.1016/j.eurpolymj.2015.03.015

V. Bershtein, A. Fainleib, P. Yakushev, Polycyanurate-based hybrid networks and nanocomposites: structure-glass transition dynamics-dynamic heterogeneity-properties relationships, in *Thermostable Polycyanurates*, ed. by A. Fainleib. *Thermostable Polycyanurates. Synthesis, Modification, Structure and Properties* (Nova Science Publishers, Inc., New York, 2010), pp. 195–245

T. Chaisuwan, H. Ishida, High-performance maleimide and nitrile-functionalized benzoxazines with good processibility for advanced composites applications. J. Appl. Polym. Sci. **101**, 548 (2006). https://doi.org/10.1002/app.23509

C.K. Chozhan, A. Chandramohan, M. Alagar, Development and characterization of 1,1-bis(3-methyl-4-hydroxyphenyl) cyclohexane-based benzoxazine and cyanate ester hybrid polymer matrices: thermal and morphological properties. Polym. Comp. **27**, 609 (2019). https://doi.org/10.1177/0967391119853747

A. Crawford, *Novel Cyanate Ester Blends* (University of Surrey, 2014), p. 207

A.O. Crawford, G. Cavalli, B.J. Howlin, I. Hamerton, Investigation of structure property relationships in liquid processible, solvent free, thermally stable bismaleimide-triazine (BT) resins. React. Funct. Polym. **102**, 110 (2016). https://doi.org/10.1016/j.reactfunctpolym.2016.03.002

M. Derradji, J. Wang, W. Liu, *Phthalonitrile Resins and Composites: Properties and Applications* (Elsevier, 2018), p. 391

A. Dhanya, S. Chandran, D. Mathew, C.P. Reghunadhan nair, in *High Performance Phthalonitrile Resins. Challenges and Engineering Applications* (De Gruyter, 2019). https://doi.org/10.1515/9783110643831

A. Fainleib (ed.), *Thermostable Polycyanurates. Synthesis, Modification, Structure and Properties* (Nova Science Publishers, Inc., New York, 2010)

H. Gu, C. Gao, A. Du, Y. Guo, H. Zhou, T. Zhao, N. Naik, Z. Guo, An overview of high-performance phthalonitrile resins: fabrication and electronic application. J. Mater, Chem. C **10**, 2925 (2022). https://doi.org/10.1039/D1TC05715D

Q. Guan, A. Gu, G. Liang, L. Yuan, F. Liao, Y. Gong, Curing kinetics and mechanism of novel high performance hyperbranched polysiloxane/bismaleimide/cyanate ester resins for resin transfer molding. J. Appl. Polym. Sci. **122**, 304 (2011). https://doi.org/10.1002/app.34073

I. Hamerton, *Chemistry and Technology of Cyanate Ester Resins* (Blackie Academic & Professional, Glasgow, 1994)

I. Hamerton, High-performance thermoset-thermoset polymer blends: a review of the chemistry of cyanate ester-bismaleimide blends. High. Perform. Polym. **8**, 83 (1996)

I. Hamerton, H. Herman, K.T. Rees, A. Chaplin, S.J. Shaw, Water uptake effects in resins based on alkenyl-modified cyanate ester—bismaleimide blends. Polym. Int. **50**(4), 475 (2001). https://doi.org/10.1002/pi.664

J. Hu, A. Gu, G. Liang, D. Zhuo, L. Yuan, Preparation and properties of high-performance polysilsesquioxanes/bismaleimide-triazine hybrids. J. Appl. Polym. Sci. **120**, 360–367 (2010). https://doi.org/10.1002/app

R. Iredale, C. Ward, I. Hamerton, Modern advances in bismaleimide resin technology: a 21st century perspective on the chemistry of addition polyimides. Prog. Polym. Sci. **69**, 1 (2017). https://doi.org/10.1016/j.progpolymsci.2016.12.002

H. Ishida, T. Agag (eds.), *Handbook of Benzoxazine Resins* (Elsevier, Amsterdam, 2011)

L. Jin, T. Agag, H. Ishida, Bis(benzoxazine-maleimide)s as a novel class of high performance resin: synthesis and properties. Eur. Polym. J. **46**, 354 (2010). https://doi.org/10.1016/j.eurpolymj.2009.09.013

T. Keller, Strong organic acid cured phthalonitrile resins for high-temperature applications. Polym. Preprint. **33**(1), 422 (1992)

T.M. Keller, Imide-containing phthalonitrile resin. Polymer **34**(5), 952 (1993). https://doi.org/10.1016/0032-3861(93)90213-T

T.M. Keller, D.D. Dominguez, High-temperature resorcinol-based phthalonitrile polymer. Polymer **46**(13), 4614 (2005). https://doi.org/10.1016/j.polymer.2005.03.068

T.M. Keller, J.R. Griffith, in *Resins for Aerospace*, ed. by C. A. May. The synthesis of a new class of polyphthalocyanine resins, American Chemical Society (1980), p. 25. https://doi.org/10.1021/bk-1980-0132.ch003

T.M. Keller, T.R. Price, Amine-cured bisphenol-linked phthalonitrile resins. J. Macromol. Sci. Part A—Chem. **18**(6), 931 (1982). https://doi.org/10.1080/00222338208077208

H. Kimura, K. Ohtsuka, A. Matsumoto, Curing reaction of bisphenol-A based benzoxazine with cyanate ester resin and the properties of the cured thermosetting resin. eXPRESS Polym. Let. **5**, 1113 (2011). https://doi.org/10.3144/expresspolymlett.2011.108

H. Kimura, K. Ohtsuka, A. Matsumoto, New thermosetting resin from benzoxazine and cyanate ester resin. Adv. Polym. Technol. **32**, E651 (2013). https://doi.org/10.1002/adv.21308

K.S.S. Kumar, C.P.R. Nair, R. Sadhana, K.N. Ninan, Benzoxazine–bismaleimide blends: curing and thermal properties. Eur. Polym. J. **43**, 5084 (2007). https://doi.org/10.1016/j.eurpolymj.2007.09.012

K.S.S. Kumar, C.P.R. Nair, K.N. Ninan, Investigations on the cure chemistry and polymer properties of benzoxazine–cyanate ester blends. Eur. Polym. J. **45**(2), 494 (2009). https://doi.org/10.1016/j.eurpolymj.2008.11.001

M. Laskoski, T.M. Keller, S.B. Qadri, Direct conversion of highly aromatic phthalonitrile thermosetting resins into carbon nanotube containing solids. Polymer **48**(26), 7484 (2007). https://doi.org/10.1016/j.polymer.2007.11.003

M. Laskoski, A. Neal, T.M. Keller, D. Dominguez, C.A. Klug, A.P. Saab, Improved synthesis of oligomeric phthalonitriles and studies designed for low temperature cure. J. Polym. Sci. Polym. Chem. **52**(12), 1662 (2014). https://doi.org/10.1002/pola.27161

M. Laskoski, M.B. Schear, A. Neal, D.D. Dominguez, H.L. Ricks-Laskoski, J. Hervey, T.M. Keller, Improved synthesis and properties of aryl ether-based oligomeric phthalonitrile resins and polymers. Polymer **67**, 185 (2015). https://doi.org/10.1016/j.polymer.2015.04.071

Y.-L. Liu, J.-M. Yu, Cocuring behaviors of benzoxazine and maleimide derivatives and the thermal properties of the cured products. J. Polym. Sci. Part A: Polym. Chem. **44**, 1890 (2006). https://doi.org/10.1002/pola.21290

V.P. McConnell, Resins for the hot zone, part II: BMIs, CEs, benzoxazines and phthalonitriles high-perform. Comp. **9**, 49 (2009)

Z. Ren, Y. Cheng, L. Kong, T. Qi, F. Xiao, High glass transition temperature bismaleimide-triazine resins based on soluble amorphous bismaleimide monomer. J. Appl. Polym. Sci. **133**, 42882 (2016). https://doi.org/10.1002/app.42882

S. Rimdusit, Ch. Jubsilp, S. Tiptipakorn, *Alloys and Composites of Polybenzoxazines. Properties and Applications* (Springer, 2013). https://doi.org/10.1007/978-981-4451-76-5

S.B. Sastri, J.P. Armistead, T.M. Keller, Phthalonitrile-carbon fiber composites. Polym. Compos. **16**(6), 816 (1996). https://doi.org/10.1002/pc.10674

S.B. Sastri, J.P. Armistead, T.M. Keller, U. Sorathia, Phthalonitrile-glass fabric composites. Polym. Compos. **18**(1), 48 (1997). https://doi.org/10.1002/pc.10260

H. Stenzenberger, Chemistry and properties of addition polyimides. In: *Polyimides*, ed. by D. Wilson, H.D. Stenzenberger, P.M. Hergenrother (Blackie & Son, Glasgow and London, 1990), p. 79. https://doi.org/10.1007/978-94-010-9661-4

T. Takeichi, Y. Saito, T. Agag, H. Muto, T. Kawauchi, High-performance polymer alloys of polybenzoxazine and bismaleimide. Polymer **49**, 1173 (2008). https://doi.org/10.1016/j.polymer.2008.01.041

Y. Wang, K. Kou, L. Zhuo, H. Chen, Y. Zhang, G. Wu, Thermal, mechanical and dielectric properties of BMI modified by the bisallyl benzoxazine. J. Polym. Res. **22**, 51 (2015a). https://doi.org/10.1007/s10965-015-0695-5

Y. Wang, K. Kou, G. Wu, L. Zhuo, J. Li, Y. Zhang, The curing reaction of benzoxazine with bismaleimide/cyanate ester resin and the properties of the terpolymer. Polymer **77**, 354 (2015b). https://doi.org/10.1016/j.polymer.2015.09.059

Y. Wang, K. Kou, G. Wu, A. Feng, L. Zhuo, The effect of bis allyl benzoxazine on the thermal, mechanical and dielectric properties of bismaleimide–cyanate blend polymers. RSC Adv. **5**, 58821 (2015c). https://doi.org/10.1039/C5RA07606D

M.W. Wang, R.J. Jeng, C.H. Lin, Origin of the rapid trimerization of cyanate ester in a benzoxazine/cyanate ester blend. Macromolecules **48**, 2417 (2015). https://doi.org/10.1021/acs.macromol.5b00334

G. Wu, K. Kou, N. Li, H. Shi, M. Chao, Electrically conductive adhesive based on bismaleimide-triazine resin filled with microcoiled carbon fibers. J. Appl. Polym. Sci. **128**(2), 1164 (2013). https://doi.org/10.1002/app.38348

G. Wu, Y. Cheng, K. Wang, Y. Wang, A. Feng, Fabrication and characterization of OMMt/BMI/CE composites with low dielectric properties and high thermal stability for electronic packaging. J. Mater. Sci.: Mater. Electron. **27**, 5592 (2016). https://doi.org/10.1007/s10854-016-4464-y

J. Yang, X. Mao, L. Du, B. Wu, F. Zhang, W. Hu, X. Tang, Thermally stabilized bismaleimide–triazine resin composites for 10-GHz level high-frequency application. High. Perform. Polym. **30**(7), 833 (2018). https://doi.org/10.1177/0954008317732396

L. Yuan, H. Yan, Y. Jia, T. Li, Microwave curing bismaleimide-triazine composites filled with micro-silica and nano-silica. J. Reinf. Plast. Compos. **33**, 1743 (2014). https://doi.org/10.1177/0731684414545142

X. Zeng, S. Yu, R. Sun, Thermal behavior and dielectric property analysis of boron nitride-filled bismaleimide-triazine resin composites. J. Appl. Polym. Sci. **128**, 1353 (2013). https://doi.org/10.1002/app.38276

Part I
Recent Advances in Studying Cyanate Ester Resin-Based Nanocomposites

Chapter 2
CER/POSS Nanocomposites

Abstract The results of physical studies of the structure, dynamics, mechanical, and thermal properties of CER/epoxy-POSS nanocomposites based on CE monomers (DCBE, DCBA, 6F-DCBA) or oligomer (PT-30) published during the last decade are presented. The nanoparticle content in the nanocomposites varied from 0.01 wt.% to 10 wt.%, and their analysis was performed by TEM, EDXS, mid-IR, far-IR, XRD, SAXS, DSC, DMA, CRS, and TGA methods. The effective molecular dispersity of nanoparticles in the amorphous matrix and the most uniform POSS distribution in the composite nanovolumes were achieved only at their contents < 1 wt.%. This and the covalent "embedding" of POSS units into the matrix led to significant suppression of matrix (cycles) dynamics and to a significant improvement in the thermal and mechanical properties of the nanocomposites. The extremal dependencies of T_g on POSS content, with its increase from 240–250° to 280–300 °C at low POSS contents and the appearance of additional interfacial dynamics transition at 375 °C, and the dynamic heterogeneity around T_g were found. The superiority of oligomer-based nanocomposites with T_g ~ 400 °C, total thermal stability up to this temperature, regardless of the nature of the environment, and the increase of the modulus from ~ 2 GPa to ~ 4 GPa at 20 °C and from 0.1 GPa to 2 GPa at 300 °C are shown.

2.1 Structure of the Nanocomposites Derived from Dicyanate Ester of Bisphenol *E* (DCBE) Monomer and PT-30 Oligomer

Early publications on this type of nanocomposites are discussed in the reviews indicated in Introductory Chap. 1. Last decade, Bershtein, Fainleib, et al. (Baikova et al. 2016; Bershtein et al. 2015, 2016, 2020; Starostenko et al. 2012, 2014) have synthesized and characterized in more detail the nanostructure, dynamics, and thermal/relaxation/elastic properties of a series of hybrid Cyanate Ester Resin (CER, Polycyanurate)-epoxy cyclohexyl-functionalized Polyhedral Oligomeric Silsesquioxane (CER/ECH-POSS) nanocomposites with the ECH-POSS contents varying from 0.01 to 10 wt. %.

V. A. Bershtein and P. N. Yakushev, *High-Temperature Polymer Nanocomposites Based on Heterocyclic Networks from Nitrile Monomers*, Springer Series in Materials Science 334, https://doi.org/10.1007/978-3-031-32943-2_2

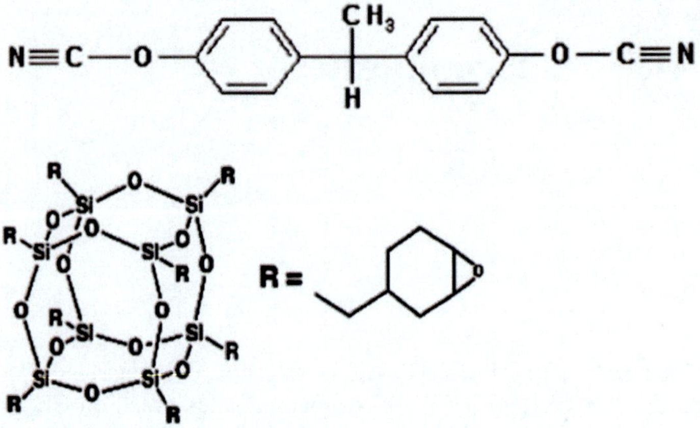

Fig. 2.1 DCBE monomer and ECH-POSS nanoparticle formulas

The main focus was on the study of nanocomposites synthesized from dicyanate ester of bisphenol *E* (DCBE) and ECH-POSS nanoparticles of about 1.5 nm in size, the schemes of which are shown in Fig. 2.1.

Additionally, CER/ECH-POSS nanocomposites derived from the other cyanated monomers and oligomer with different molecular structures were also synthesized and characterized (see below).

The overall goal of the comprehensive physical study (Baikova et al. 2016; Bershtein et al. 2015, 2016, 2020; Starostenko et al. 2012, 2014) was to find the relationship between the structure, matrix dynamics, and thermal, relaxation, and elastic properties of the above nanocomposites. Within the framework of this research such methods as transmission electron microscopy (TEM), Fourier transform infrared (FTIR) spectroscopy, far-infrared (Far-IR) spectroscopy, small-angle *X*-ray scattering (SAXS) technique, energy dispersive *X*-ray spectroscopy (EDXS), dynamic mechanical analysis (DMA), differential scanning calorimetry (DSC), thermogravimetric analysis (TGA), and laser-interferometric creep rate spectroscopy (CRS) (Bershtein and Yakushev 2010) technique have been used.

The curing of the DCBE/ECH-POSS mixtures was performed by heating at a rate of 0.5 °C/min from 25 to 300 °C. Under these conditions a complete polymerization ("polycyclotrimerization") with the formation of a polymer network occurred, which was proved by the disappearance of the absorption bands of the nitrile (cyanate) groups at 2237 and 2266 cm^{-1} in the infrared spectra.

Several chemical reactions occurred at high temperatures between cyanate groups and also between cyanate and epoxy groups, as it is presented schematically in Fig. 2.2. As it was clear from FTIR spectra, they consist of the polymerization of the DCBE monomer into a polycyanurate network containing triazine heterocycles (reaction 1); reaction 2 of some cyanate groups with the epoxy groups of ECH-POSS nanoparticles to form oxazoline cycles and partial isomerization of the latter into oxazolidinone cycles (reaction 3). In addition, some of the formed triazine cycles could also react with the epoxy groups to form the isocyanurate cycles (reaction 4), which could also chemically interact with the epoxy groups of ECH-POSS to form the oxazolidinone cycles (reaction 5).

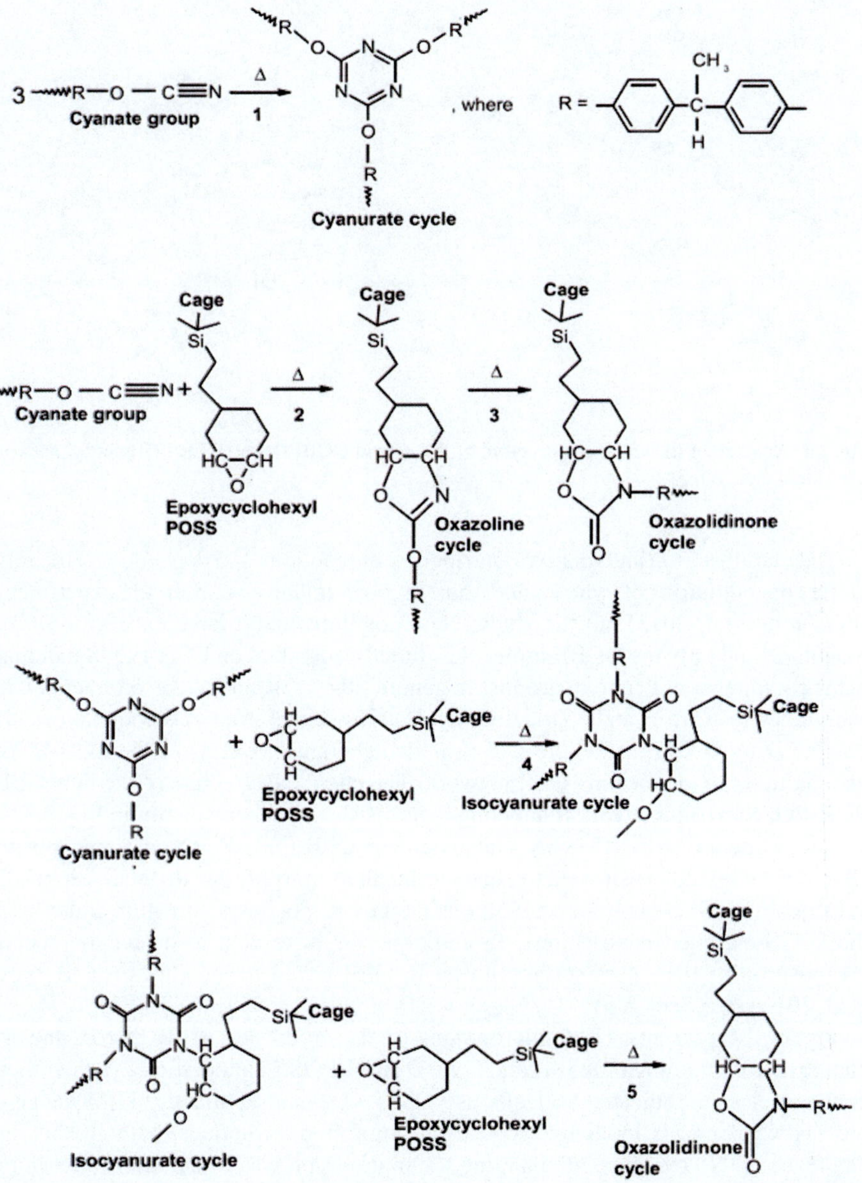

Fig. 2.2 Scheme of the possible chemical transformations in the DCBE-based CER/ECH-POSS system in the curing process (Bershtein et al. 2015)

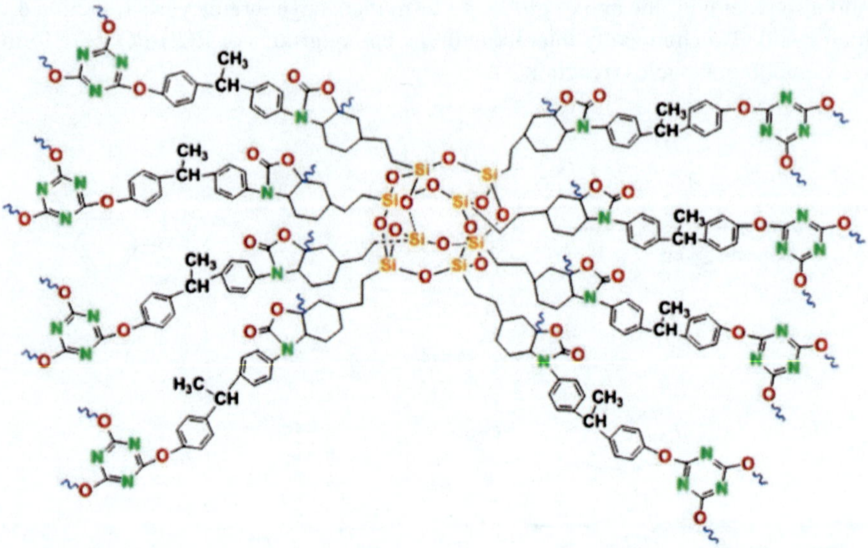

Fig. 2.3 Scheme of the chemical structure of the hybrid DCBE-based CER/ECH-POSS network (Bershtein et al. 2015)

The manifestation and relative contribution of reactions 1–5 depended on the ratio of the concentrations of cyanate and epoxy groups in the system. In the experiments (Bershtein et al. 2015), all the studied compositions had a large excess of DCBE monomer, and only the DCBE molecules directly adsorbed on ECH-POSS nanoparticles could react with epoxy groups; in general, the CER molecules interacted with each other to form cyanate (triazine) cycles. However, at any composition of the polymerizing mixture, reactions 2–5 provided chemical grafting of the ECH-POSS nanoparticles to the forming CER network. The chemical structure of the final CER/ECH-POSS hybrid network (nanocomposite) is shown schematically in Fig. 2.3.

The joint TEM/EDXS/SAXS analysis of CER/ECH-POSS nanocomposites (Bershtein et al. 2015) allowed to characterize their morphology in detail, especially to estimate the Si content in nanovolumes of the composites both within and outside the POSS-enriched nanoregions. In addition, we were able to detect the special composition of the 1-μm surface layers in CER-based nanocomposites (Baikova et al. 2016) (see Sect. 3.6).

Figure 2.4 presents TEM micrographs of the cured neat CER matrix and the composites containing 0.025, 1, and 2 wt.% ECH-POSS nanoparticles (below these nanoparticles are indicated basically as POSS). One can see that the TEM image of the neat CER matrix looks like typical for amorphous structures, without showing any features. No evidence of changing the micrograph was found after introducing 0.025 wt.% or 1 wt.% POSS nanoparticles. It follows that in these cases, the studied nanocomposites are characterized by the apparently molecular level of POSS nanoparticles distribution in the matrix, i.e., by the absence of their clustering. Under

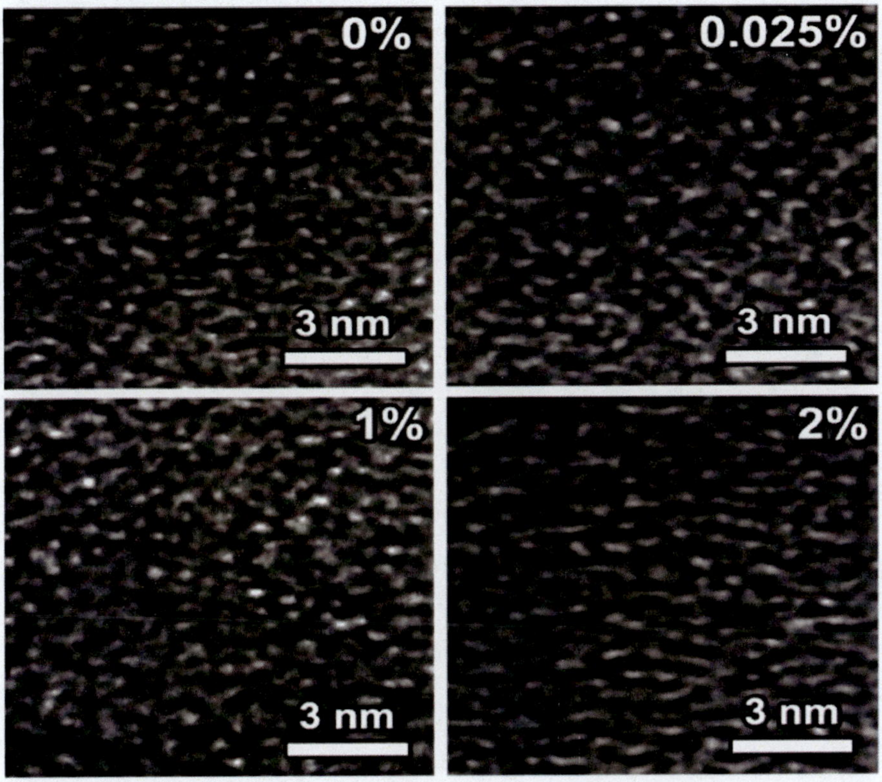

Fig. 2.4 TEM micrographs obtained for neat DCBE-based CER matrix and CER/ECH-POSS nanocomposites with 0.025, 1, and 2 wt. % nanoparticles (Bershtein et al. 2015)

these conditions, each POSS nanoparticle forms several chemical bonds with the matrix.

Such an effective dispersion of POSS nanoparticles could be explained by the rapid mixing of POSS with the monomer and the rapid chemical reaction between nitrile and epoxy groups. Naturally, then, the POSS molecules "bound" in this way were no longer free to self-aggregate.

In contrast to this, the micrograph presented in Fig. 2.4 makes it possible to recognize some structural elements in a nanocomposite containing 2 wt.% POSS. In this case, it was possible to distinguish both a larger number and an elongated shape of some nanoinclusions, which were either nanorod- or, possibly, platelet-like domains. Already these elements can be presumably associated with a certain interaction of POSS nanoparticles. TEM of nanocomposite samples containing 5 or 10 wt.% POSS already quite clearly showed the emergence of dark spherical spots of ca. 50–100 nm in size in the TEM images, i.e., presumably relatively large aggregates of POSS nanoparticles (Fig. 2.5).

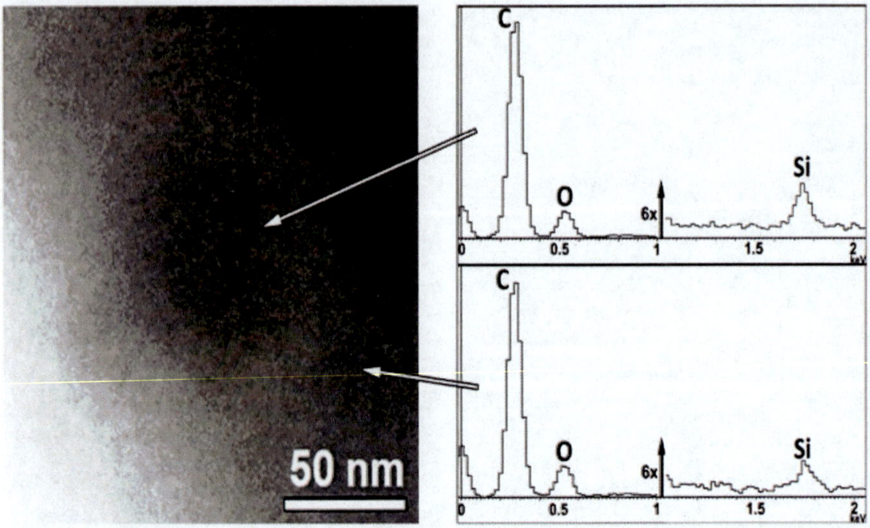

Fig. 2.5 TEM image of a nanoaggregate in the DCBE-based CER/ECH-POSS (10 wt.%) nanocomposite, and two EDX spectra showing the comparative Si/O/C atomic content ratio in the "dark" region of about 70–100 nm in size and in the surrounding more "light" matrix. The "darker" region corresponds to POSS-enriched/polymer matrix nanovolume with the higher Si content than that in the more "light" surrounding matrix (Bershtein et al. 2015)

The assumption that this is indeed a POSS-enriched inclusion was confirmed experimentally by the EDXS method when comparing EDX spectra characterizing the elemental composition inside the indicated dark spot and outside it in the surrounding matrix. Figure 2.5 compares these spectra. It may be seen that at the invariable C and O atoms contents in both spectra Si atom content is higher inside the dark region than that in the surrounding matrix. This difference, however, is much less than could be expected. This result indicates the absence of formation of real individual POSS nanophases. Additionally, wide-angle *X*-ray diffraction experiments (Bershtein et al. 2015) also indicated the absence of crystalline POSS aggregates in these nanocomposites. It means that the POSS-enriched nanoregions incorporated also CER matrix. Generally, at increased nanofiller contents one could expect the concurrence between the chemical incorporation of the POSS units into the forming polymer network and their aggregation. It is noteworthy that Liang et al. (2006) have also revealed, using TEM/SEM/SANS analysis, rather similar structural manifestations in CER-based nanocomposites.

EDXS also allowed us to obtain interesting information about the distribution pattern of POSS nanoparticles in the nanocomposites under discussion. In these experiments, the local content of Si atoms was measured at 50 different points for each nanocomposite (Bershtein et al. 2015) In this case, the electron beam was focused on sites with a diameter of about 10 nm. The obtained data made it possible to construct histograms of the local Si atom content, which were characteristic of the degree of homogeneity of the POSS nanoparticles distribution in the matrix nanovolumes.

Figure 2.6 shows the histograms of local Si content distribution obtained for four nanocomposites containing 0.5, 2, 5, or 10 wt.% ECH-POSS nanoparticles. The dashed lines correspond to the average values of the Si content in the composites. In general, one can see that POSS particles are present in all analyzed nanovolumes, and their heterogeneous distribution in the matrix is typical for all four nanocomposites; however, the shapes of the obtained histograms differ significantly. Really, at 0.5 wt.% ECH-POSS in the composite, the average Si content was recorded in 48% of the points; most of the other points also showed similar results, and only a few points exhibited significantly decreased or increased Si content (Fig. 2.6a). The histogram of Si (POSS) distribution changed at 2 wt.% ECH-POSS (Fig. 2.6b), in particular, the spots with sharply decreased Si contents disappeared. However, the histogram maximum coincided with the average Si content as well, and the absolute majority of the spots showed Si contents within the range between 0.1 and 0.2 of Si atom percent. Meanwhile, at 5 and 10 wt.% ECH-POSS the histograms sharply changed distinctly indicating an increase in nanoheterogeneity of POSS distribution.

At 5 wt.% ECH-POSS, the average Si content was observed in only 15% of spots; the histogram maximum was far from the average Si content and the Si content varied in several times, with significantly increased and decreased values (Fig. 2.6c). Finally, at 10 wt.% ECH-POSS, again a specific histogram was observed when nanodomains with several different Si contents were registered, including sharply POSS-enriched and POSS-depleted nanodomains (Fig. 2.6d). Thus, the most degree of structural heterogeneity was observed in the latter case, and the common trend herein was as follows: the heterogeneity of POSS distribution became more pronounced with increasing their content. Therefore, the most uniform POSS distribution in the nanocomposite should be expected, presumably, at the ultra-low contents of these nanoparticles. Unfortunately, histograms of nanocomposites with ultra-low nanoparticle content could not be obtained for methodological reasons.

Finally, additional information about the structure of the nanocomposites in question was obtained using the small-angle X-ray scattering (SAXS) method (Bershtein et al. 2015). It is known that the SAXS plot is sensitive to local variations in the electron density caused by the presence of nanodomains that have a higher or lower density than the average material density. This allows estimating the structural nanoheterogeneity of the material with characteristic spatial dimensions (Bragg periodicity $2\pi/q$, where q is wave vector). Figure 2.7 shows SAXS images, the experimental $I(q)$ versus q scattering curves, obtained over q range from 0.2 to 1.3 nm^{-1}, for the neat cured CER matrix and the nanocomposites with 0.025, 1, and 10 wt.% POSS.

One can see that the scattering curve of the neat CER matrix is characterized by the absence of peaks that indicates the lack of structural nanoheterogeneity. At the same time, a common feature of SAXS curves of hybrid nanocomposites is the presence of wide low-intensity "maxima" (which look rather like shoulders) in the interval q from 0.23 nm^{-1} to 0.69 nm^{-1}. This effect appears even at the minimum nanoparticle content (0.025 wt.%) in the nanocomposite. This result confirms the data on the

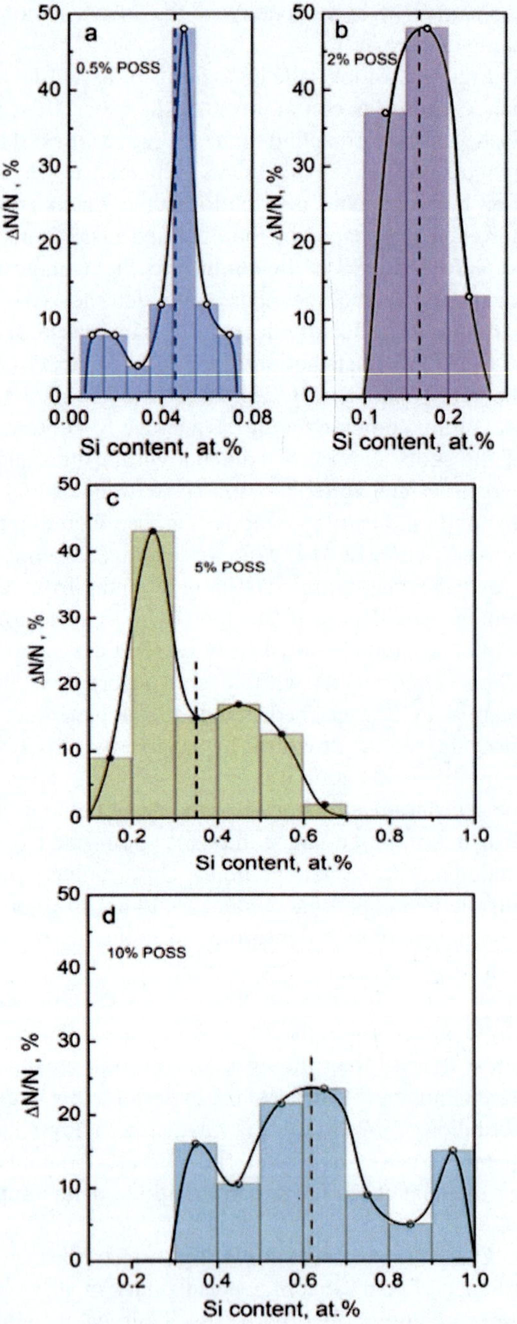

Fig. 2.6 Histograms of local silicon content distribution in the DCBE-based CER/ECH-POSS (0.5 wt.%), CER/ECH-POSS (2 wt.%),), CER/ECH-POSS (5 wt.%), and CER/ECH-POSS (10 wt.%) nanocomposites (Bershtein et al. 2015)

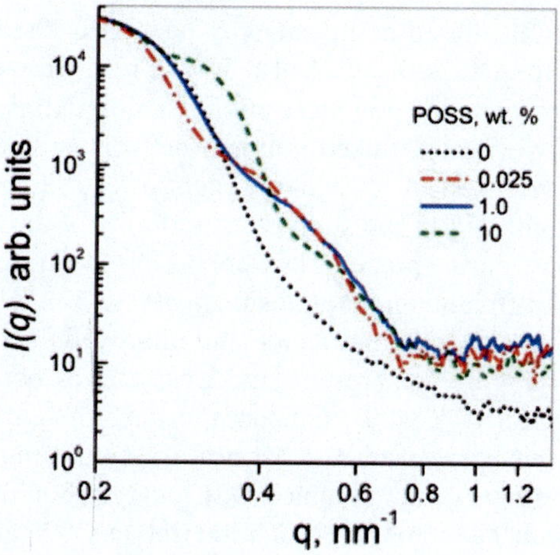

Fig. 2.7 SAXS curves obtained for neat DCBE-based CER matrix and CER/ECH-POSS nanocomposites with 0.025 wt.%, 1 wt.%, and 10 wt. % nanofillers (Bershtein et al. 2015)

emergence of structural heterogeneity in the investigated nanocomposites. However, here also each composition has its own peculiarities. Thus, for a nanocomposite with 0.025 wt.% POSS, the scattering "maximum" is localized in the interval q from 0.3 to 0.68 nm^{-1} corresponding to Bragg's periodicities of 16 ± 5 nm. This result does not generally contradict the TEM data, which suggest a mainly quasi-periodic spatial distribution of individual POSS nanoparticles in the amorphous matrix.

A scattering "maximum" for the composite with 1 wt.% POSS is characterized by the q values ranging from 0.39 nm^{-1} to 0.70 nm^{-1}, i.e., with the lesser Bragg's periodicity values, of 12 ± 3 nm. Of special interest, the composite with the highest POSS loading (10 wt.%) containing, according to TEM estimation, not only molecularly dispersed POSS units but also POSS-matrix aggregated nanodomains, manifested two distinct broad scattering "maxima" at q = 0.23 – 0.42 nm^{-1} and 0.42 – 0.67 nm^{-1} corresponding to Bragg's periodicities of 12 ± 3 nm and 21 ± 6 nm (Fig. 2.7). It could be suggested that the first scattering "maximum" may be associated with the relatively homogeneous distribution of a part of POSS units within the matrix, whereas the "maximum" with the periodicity of 21 ± 6 nm can be attributed to the presence of POSS-matrix aggregates in the CER/ECH-POSS (10 wt.%) nanocomposite. As follows from the Hosemann-Bagchi conception (Hosemann and Bagchi 1962), the appearance of a scattering maximum requires the sizes of "nanoinclusions" (domains with anomalous densities) not less than 3–4 values of periodicity calculated from the position of the corresponding peak. It means that the size of such POSS-enriched domains in the CER matrix must be not less than 50–60 nm that is in accordance with the TEM data (Fig. 2.5).

Bershtein et al. (2020) have performed also a detailed analysis of the structure of a series of multi-functional PT-30 oligomer-based CER/ECH-POSS nanocomposites with 0.025 to 10 wt.% nanofillers. The formula of the PT-30 oligomer and the scheme of the densely cross-linked PT-30-based matrix network are shown in Fig. 2.8. Initially, the neat and filled pre-polymers were cured as in the case of

DCBE-based composites by heating in the temperature range from 25 to 300 °C, with the heating rate of 0.5 °C min^{-1}. However, in this case it was not possible to complete the polymerization reaction in this regime. Apparently, the formation of a perfect cross-linked polymer network in this case was hindered by such factors as steric obstacles, higher content of nitrile groups (compared to DCBE), as well as high rigidity and suppressed mobility of the forming network. Therefore, additional post-curing procedures were performed in Bershtein et al. (2020) by heating for 0.5 h at different temperatures from 300 to 420 °C. This resulted in a much higher degree of matrix polymerization and improved the characteristics of the nanocomposites.

Figure 2.9 presents the DSC curves obtained for the PT-30-based CER/ECH-POSS (0.5 wt.%) nanocomposite films cured by heating from 20 to 300 °C and after post-curing for 0.5 h at different temperatures from 300 to 420 °C. Several features of the obtained DSC curves should be noted. First, the heat capacity step, with $T_g = 264$ °C, was observed only for the composite with initially polymerized, highly defective matrix network; in the case of post-cured samples such steps were absent in the DSC curves. Secondly, exotherms of the post-polymerization process were generally observed starting from temperatures of 360–370 °C for both initially cured samples and the samples post-cured at 300, 320, or 340 °C. The exotherm area decreased with increasing post-curing temperature and the exotherm completely disappeared after post-curing at 360 °C. It was natural to suppose that the exotherms of the post-polymerization process occurred due to arising some network mobility at 360–370 °C. This "unfreezing" of mobility, providing some possibilities for continuation of reaction of polymerization, may be considered as a slight manifestation of "glass transition" in this polymer network; similar connection between annealing temperature and polymerization degree (cross-linking density, T_g) is well known for polymer networks, in particular for epoxies. As seen from Fig. 2.9, this specific temperature may increase after post-curing from 360–370 °C to 410 °C and then decreases due to starting degradation after annealing at 380 or 420 °C. And, third,

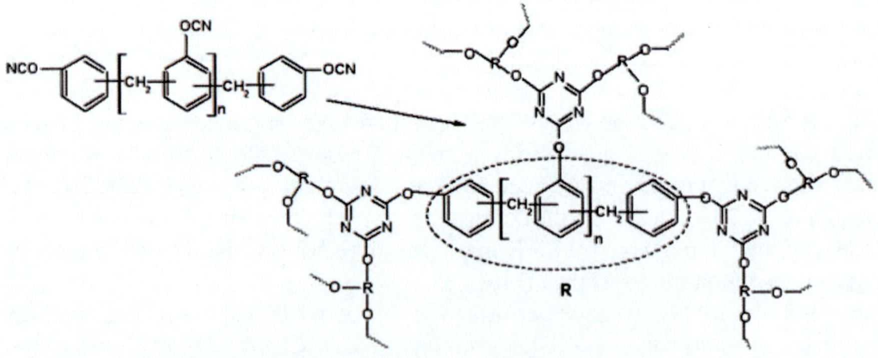

Fig. 2.8 Formula of PT-30 oligomer ($n \approx 2$) and the scheme of densely cross-linked PT-30-based CER network (Bershtein et al. 2020).

the stronger exothermic process associated with the thermo-oxidative degradation occurred from temperatures of about 500 °C (see below).

Figure 2.10 shows IR spectra of the PT-30-based CER/ECH-POSS (0.5 wt.%) nanocomposite films, cured by heating from 20 to 300 °C and post-cured for 0.5 h at 360 °C, within the regions of absorption bands of reactive cyanate (–O–C≡N) groups (2250, 2209 cm^{-1}). The 2955 cm^{-1} absorption band was used as an internal standard to control changes in the effective thickness of the sample.

A temperature of 360 °C was optimal for the post-polymerization process (see DSC curves in Fig. 2.9). The IR spectra indicate the following. Firstly, polymerization of PT-30 oligomer occurs only partially upon heating to 300 °C: in the spectrum, intense absorption bands of 2250 and 2209 cm^{-1} are preserved. Secondly, post-curing at 360 °C leads to a complete disappearance of the 2209 cm^{-1} absorption band, but only a two-fold decrease in the intensity of the 2250 cm^{-1} band. This indicates

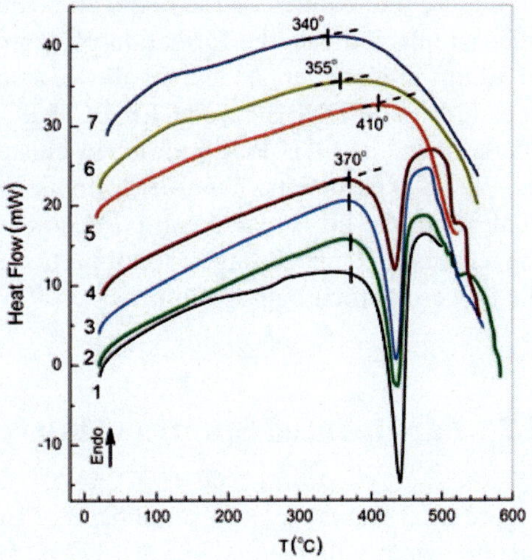

Fig. 2.9 DSC curves obtained for the PT-30-based CER/ECH-POSS (0.5 wt.%) nanocomposite films cured by heating from 20 to 300 °C (1) and after additional post-curing for 0.5 h at 300 (2), 320 (3), 340 (4), 360 (5), 380 (6), and 420 °C (7) (Bershtein et al. 2020)

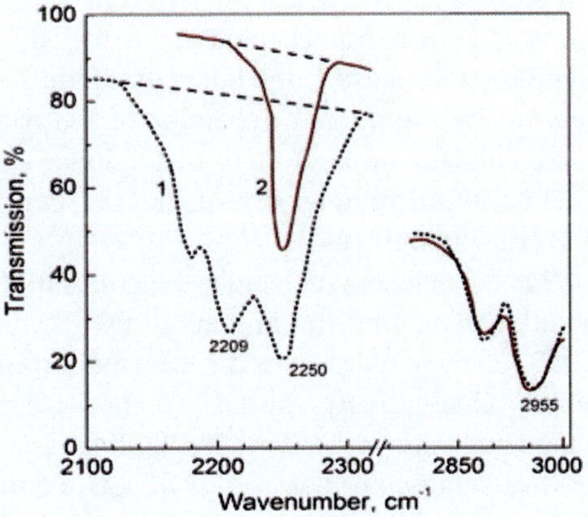

Fig. 2.10 IR spectra of the PT-30-based CER/ECH-POSS (0.5 wt.%) nanocomposite films cured by heating from 20 to 300 °C (1) and post-cured for 0.5 h at 360 °C (2). The spectra are shown within the spectral regions of absorption bands of reactive cyanate (–O–C≡N) groups (2250, 2209 cm^{-1}) and internal standard at 2955 cm^{-1} (Bershtein et al. 2020)

additional cross-linking of the matrix, but still an impossibility to complete the polymerization process. This is obviously due to the greater difficulties (steric hindrances) in cross-linking of multi-functional CER oligomers as compared to bifunctional CER monomers.

Figure 2.11 shows TEM micrographs of several PT-30-based nanocomposites. It is clear that the effective dispersion of ECH-POSS nanoparticles could occur during their rapid reactive mixing with the oligomer and their covalent "embedding" into the forming polymer network through the reaction of cyanate and epoxy groups. It is under these conditions that ECH-POSS nanoparticles prove incapable of forming aggregates (nanoclusters). TEM images of the structure of nanocomposites with 0.05 wt. % and 0.5 wt. % ECH-POSS were typical for amorphous structures, in which there are no manifestations of nanoparticle aggregation. This undoubtedly indicates a mainly molecular level of dispersity of nanoparticles in the matrix.

In contrast, at higher nanoparticle content in the composites, 2 wt. % and especially 10 wt. %, the features of the nanocomposite structure become clearly visible: the micrographs indicate the formation of aggregates-dark spherical spots of about 20 to 30 nm in diameter. As shown above, a similar result was observed in Bershtein et al. (2015) for DCBE-based CER/ECH-POSS nanocomposites; the nature of these "dark spots" as ECH-POSS-enriched clusters in nanocomposites was revealed in the EDXS experiments. Thus, the formation of nanoclusters at elevated nanoparticle contents in the investigated composites is actually due to the competition between the reaction of "embedding" of nanoparticles into the forming polymer network and the process of their aggregation.

2.2 Far-Infrared Spectra (Matrix Dynamics)

Figure 2.12 shows the absorption band of ring vibrations at 76 cm^{-1} and two overlapping bands with the maxima at 187 cm^{-1} and 230 cm^{-1} in the spectrum of a neat DCBE-based CER network. Introduction of 0.025 wt.% POSS units exerts an essential influence on these three absorption bands, i.e., on CER network dynamics; far-IR spectra in this region characterize matrix dynamics (Bershtein and Ryzhov 1994). A significant decrease in the intensity of the 76 cm^{-1} band indicates a certain suppression of the vibrational dynamics of the triazine and benzene circles. In addition, some changes in the system of torsional skeletal vibrations are observed, namely, a decrease in the intensity of the 187 cm^{-1} band and a shift of the maximum at 230 cm^{-1} to 220 cm^{-1}.

Such changes in molecular dynamics of the matrix caused by the chemical incorporation of an ultra-low amount of POSS units became possible, obviously, only as a result of their quasi-periodic distribution at the molecular level in the nanocomposite. Qualitatively similar effects were registered in the far-IR spectra of the nanocomposite with 0.5 wt.% POSS units.

It is significant that at high POSS loading of the CER matrix (10 wt.%) quite different spectral changes were registered (Bershtein et al. 2015). In this case, in the

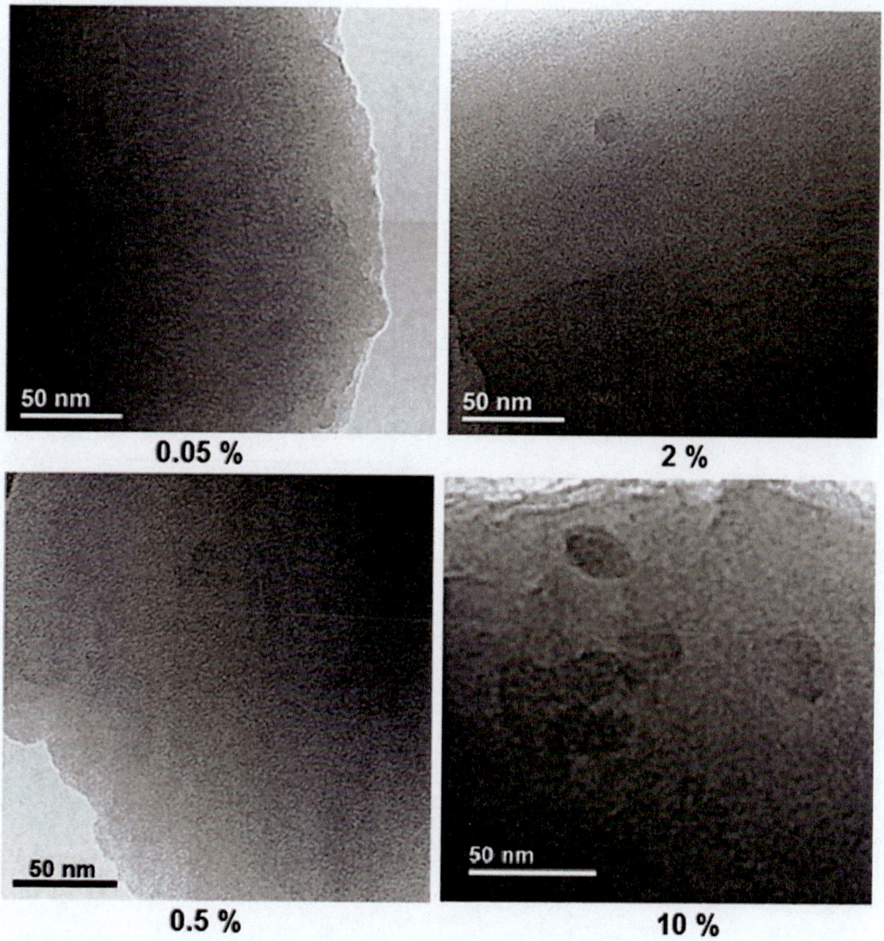

Fig. 2.11 TEM micrographs obtained for the PT-30-based CER/ECH-POSS nanocomposites with 0.05, 0.5, 2, and 10 wt.% ECH-POSS nanoparticles (Bershtein et al. 2020)

presence of POSS-containing aggregates in the nanocomposite structure, there was no notable effect of disturbing (suppression) of motion in the matrix by nanofiller (there was no change in bands at 76 cm^{-1} and 187 cm^{-1}), and the 230 cm^{-1} band was transformed into a doublet at 220 cm^{-1} and 230 cm^{-1}. Such a "bifurcation" of the absorption band could presumably be attributed to dynamics in POSS-enriched and sharply POSS-depleted nanodomains.

The additional information on matrix dynamics was obtained from the far-IR spectra of the PT-30-based CER/ECH-POSS materials (Bershtein et al. 2020) Fig. 2.13 shows the broad absorption band at 60–130 cm^{-1} with the maximum at 85 cm^{-1}. One can see that post-curing and especially post-curing in combination with the chemical embedding 0.5 wt.% ECH-POSS nanoparticles into the matrix resulted,

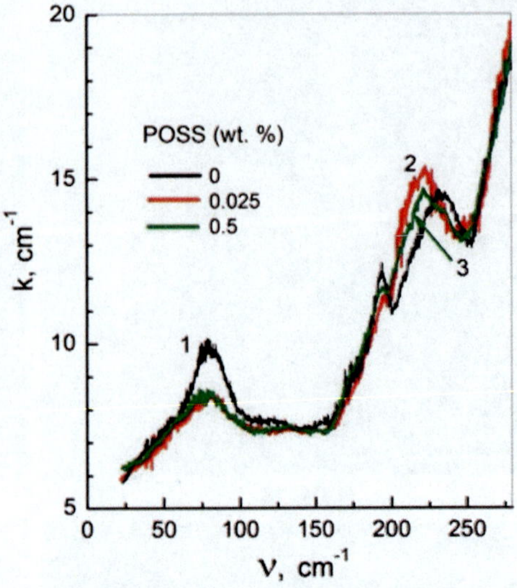

Fig. 2.12 Far-IR spectra of neat DCBE-based CER matrix and the CER/ECH-POSS nanocomposites with 0.025 wt.% and 0.5 wt.% nanofillers (Bershtein et al. 2015)

certainly, in some suppression of network (circles) dynamics. It manifested itself in decreasing band intensity as well as its small displacement to higher wavenumbers in the spectrum of the nanocomposite. These effects may be related to the well-known phenomenon of "constrained dynamics" (see, e.g., reviews (Bershtein and Yakushev 2010; Giannelis et al. 1999)).

It is shown below that the structural TEM/EDXS/SAXS data discussed and the far-IR spectra are in agreement with the mechanical and thermal properties of the nanocomposites. This is evidence in favor of the obvious correlation between their structure, dynamics, and properties. Interestingly, a general rule is observed in all cases: depending on the nanoparticle content, their ambivalent influence on the

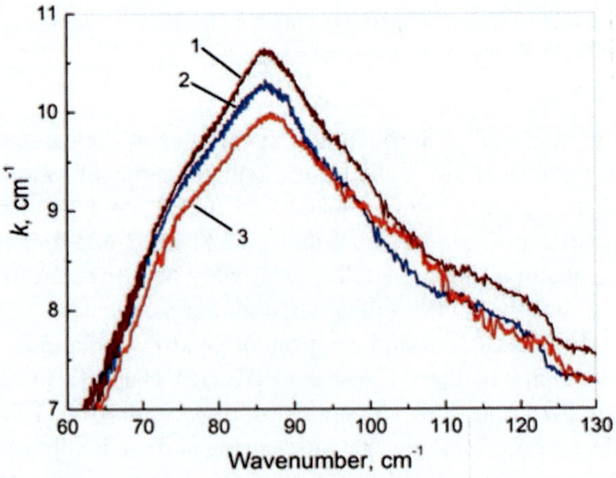

Fig. 2.13 Far-IR spectra: the absorption band of small-angle circle vibrations in the polymer networks of (1) cured unfilled PT-30-based CER matrix, (2) this matrix post-cured for 0.5 h at 360 °C, and (3) for the post-cured PT-30-based CER/ECH-POSS (0.5%) nanocomposite (Bershtein et al. 2020)

nanocomposite dynamics and properties is registered. As shown in the following sections of Part I, the maximum positive effects are observed at ultra-low contents of the introduced nanoparticles.

2.3 Glass Transition, Dynamic Mechanical Analysis

Initially, it was shown in the DSC experiments that at such a small ECH-POSS nanoparticle content as 0.025–0.1 wt.%, cardinal changes in the glass transition characteristics of the DCBE-based CER matrix were observed. In this case, the main transition appears at a significantly higher temperature and a second, much higher-temperature transition appears (Bershtein et al. 2015; Starostenko et al. 2012).

Instead of a single glass transition with $T_g = 244$ °C in the neat matrix, the DSC curve recorded a main transition with $T_g = 275$ °C and a less intense transition at $T_g = 375$ °C in the nanocomposite. In addition, there was an even higher, by 50° increase in the onset temperature of the main transition, $T_g{'}$, and a narrowing of the range of this transition. The higher-temperature glass transition (which was followed by the thermo-oxidative degradation process) was presumably attributed to the manifestation of dynamics in the interphase layers (strong "constrained dynamics" effect).

At the same time, the opposite tendency—a decrease in the glass transition temperatures, especially of transition onset T_g', with a broadening of the transition range—was observed at high POSS nanoparticle content in the composites; in any case, their positive influence disappeared. Such a negative effect could be explained, in particular, by a local decrease in the cross-linking density in the matrix due to the high consumption of cyanate groups for the co-reaction with nanoparticles. The above is illustrated by Fig. 2.14, which shows the dependencies of the basic glass transition temperatures in the DCBE-based CER/ECH-POSS nanocomposites on the nanoparticle content (Starostenko et al. 2012) This figure also shows the enhancement of thermal stability of nanocomposites with ultra-low nanoparticle contents at the early stage of thermal degradation, after scanning up to 400 °C at a rate of 20 °C min^{-1}.

Qualitatively similar results were obtained in this case by the DMA method. Figure 2.15 shows that on the $\tan \delta(T)$ dependence for the neat CER matrix the main peak in the glass transition has a maximum at $T_g = 248$ °C, while for the nanocomposite with 0.025 wt.% ECH-POSS nanoparticles a glass transition peak at $T_g = 265$ °C and two small overlapping peaks at 370–390 and 430 °C are observed. The latter peaks were attributed, respectively, to interfacial matrix dynamics and the onset of the thermos-oxidative degradation process. Additionally, the storage (dynamic) modulus *E*' of this nanocomposite increased five times, from 0.1 GPa to 0.5 GPa, at 250 °C. At 0.5 wt.% nanofillers, this modulus increased by 30–40% within the 20 to 200 °C range (Fig. 2.15).

Figure 2.16 shows the trend of changing the characteristic glass transition temperatures with variations in POSS nanoparticle content in the investigated series of

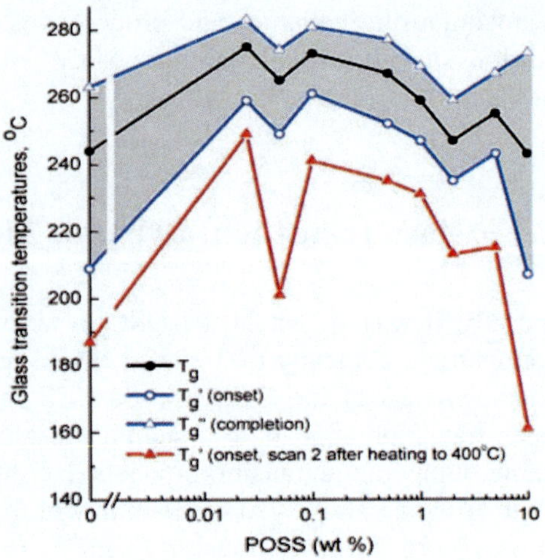

Fig. 2.14 DSC data: main glass transition temperatures of the DCBE-based CER/ECH-POSS nanocomposites as a function of nanofiller content (Starostenko et al. 2012)

nanocomposites when measured with both DSC and DMA. DMA glass transition temperatures were defined here as the temperatures of the beginning of the relaxation peaks, as maxima in the loss modulus E''(T) curves, and as maxima (T_g) in the tanδ (T) curves. Again, this figure shows a substantial T_g rise after introducing ultra-low POSS content of $0.025 - 0.05$ wt.%; the further slight changes in transition temperatures at the other low POSS contents < 1 wt.%, and basically the opposite tendency at the larger POSS loadings of 2 wt. % to 10 wt.%. At high POSS loadings, the transition onset and T_g temperatures become approximately the same as for the neat CER matrix.

Thus, the most noteworthy result of the above series of experiments should be considered the maximal positive influence on the polymer dynamics of the introduction of ultra-small amounts of 3D nanoparticles into the matrix, viz., as low as 0.025 wt. %. Actually, for matrices, linear polymers or loosely cross-linked polymer networks-full "constraining dynamics" by 3D nanoparticles can be expected only under certain conditions. This can probably occur in principle if an average interparticle distance, L, is close in value or smaller than the unperturbed dimensions of macromolecular random coils (radius of gyration R_g); this parameter for many polymers has a value of about 10 nm (Bershtein and Yakushev 2010; Giannelis et al. 1999) Therefore, a few percent loading was typically required for attaining the substantial constraining dynamics by 3D nanoparticles. In the case of semi-interpenetrating networks, such an effect was strongly pronounced at 0.25 wt. % nanoparticles (nanodiamonds) only when $L > R_g$; this result has been explained by the double covalent bonding (hybridization) between the matrix components (Bershtein et al. 2008).

Therefore, the anomalously large impact of the ultra-low ECH-POSS amounts, 0.025–0.05 wt.%, in the composites on CER glass transition dynamics could be explained, in our opinion, by the combined action of a few factors, viz.: (a) effective dispersion and relatively homogeneous distribution of ~ 1-nm size ECH-POSS

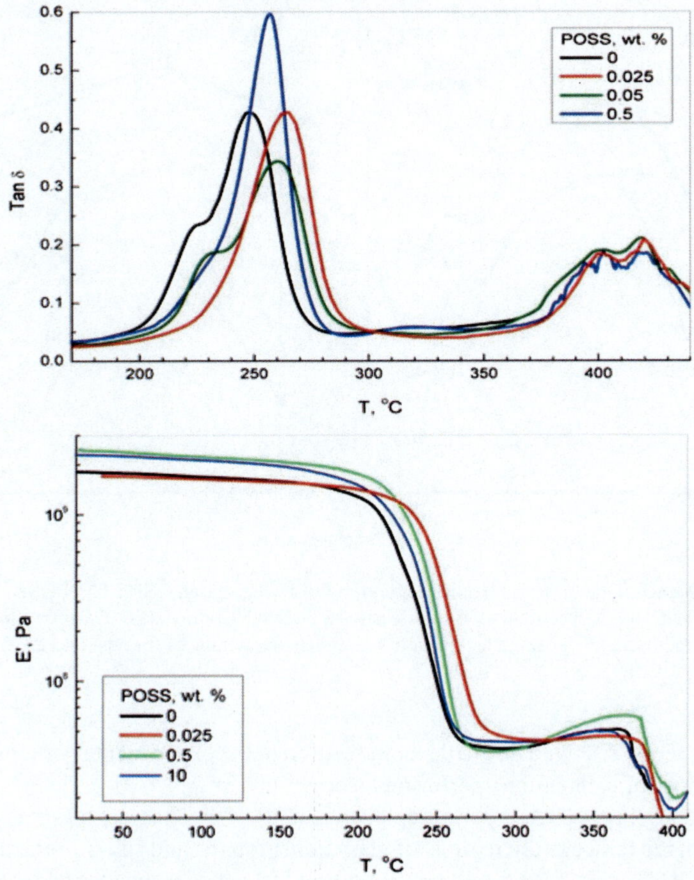

Fig. 2.15 DMA data (1 Hz): above–tanδ and below- dynamic modulus *E'* versus temperature plots obtained for the DCBE-based CER matrix and CER/ECH-POSS nanocomposites with 0.025, 0.05, 0.5, and 10 wt. % nanofillers (Starostenko et al. 2012)

molecules of cage structure (unbound "silica nanoblocks") with an extremely large specific surface area of a few thousands of m^2/g; (b) strong interphase interactions due to covalent bonding of nanoparticles to the forming matrix network; (c) due to the increased cross-linking density of the polymer matrix caused by the "embedding", additionally, of "silica nodes", and (d) finally, due to the increased impact of the embedded nanoparticles on the CER dynamics of the matrix (a strongly pronounced "constrained dynamics" effect).

Regarding the latter reason, we supposed that anomalously strong influence of ultra-low POSS loading on matrix dynamics might be caused by more long-range impact of nanoparticles on the dynamics in the densely cross-linked matrix than in the linear or loosely cross-linked polymer matrices (Starostenko et al. 2012) This problem of the anomalously large impact of the ultra-low content of nanoparticles

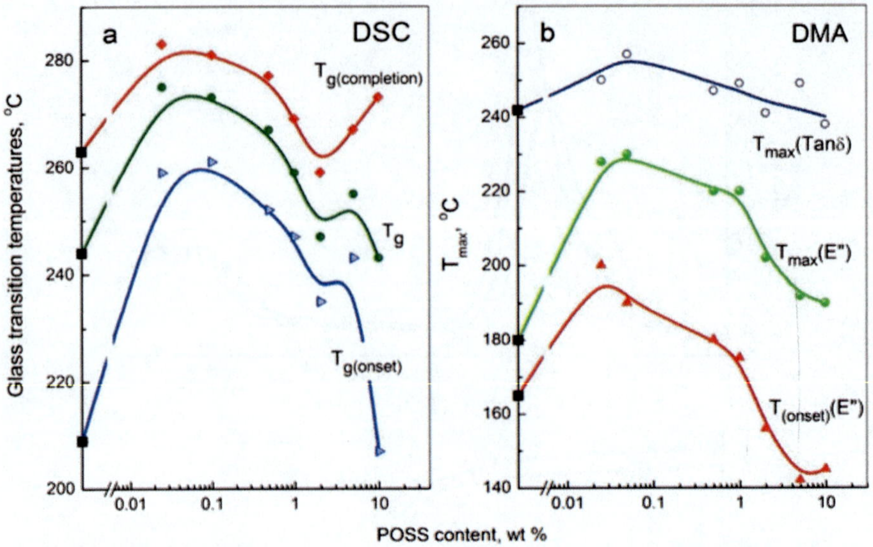

Fig. 2.16 Characteristic glass transition temperatures of DCBE-based CER/ECH-POSS nanocomposites versus ECH-POSS content plots as estimated **a** by DSC (heating at 20 °C/min) or **b** by DMA (tanδ or loss modulus E'', 1 Hz). Black squares designate temperatures for neat cured CER matrix (Bershtein et al. 2015)

was subsequently resolved due to the combination of the experimental structural data with a number of calculations performed (see Sect. 4.3).

Figure 2.17 shows the temperature dependencies of mechanical loss factor, tanδ, measured in the temperature regions of glass transition (**a**) and sub-T_g relaxation (**b**) for pristine PT-30 oligomer-based matrix and several nanocomposites based thereon, with different ECH-POSS contents, cured by heating from 20 to 300 °C.

One can see that this oligomer-based CER network manifests obvious superiority over the network synthesized from DCBE monomer. Glass transition relaxation peak of the cured PT-30-based matrix is characterized by the doublet with two T_g s, at 392 and 407 °C. The doublet character of this peak was associated probably with the post-curing process (see below) occurring, to some extent, during the DMA experiment (heating with the rate of 3 °C min^{-1}). Additionally, the network rigidity increased substantially at replacing DCBE monomer for PT-30 oligomer in the synthesis of the matrix: at 20 °C dynamic (storage) modulus E' increased from about 2 GPa to 4 GPa. At 300 °C, the modulus of the DCBE-based network was negligibly small (< 0.1 GPa) whereas $E' \approx 2$ GPa for the cured PT-30-based matrix network (Bershtein et al. 2020).

Figure 2.18 shows the glass transition temperatures, T_g *s*, as the functions of ECH-POSS content in cured nanocomposites.

From Figs. 2.17 and 2.18, one can see that covalent embedding ECH-POSS nanoparticles in the PT-30-based network resulted in a small but ambivalent (as above in the other nanocomposites) influence on both T_g and sub-T_g relaxations.

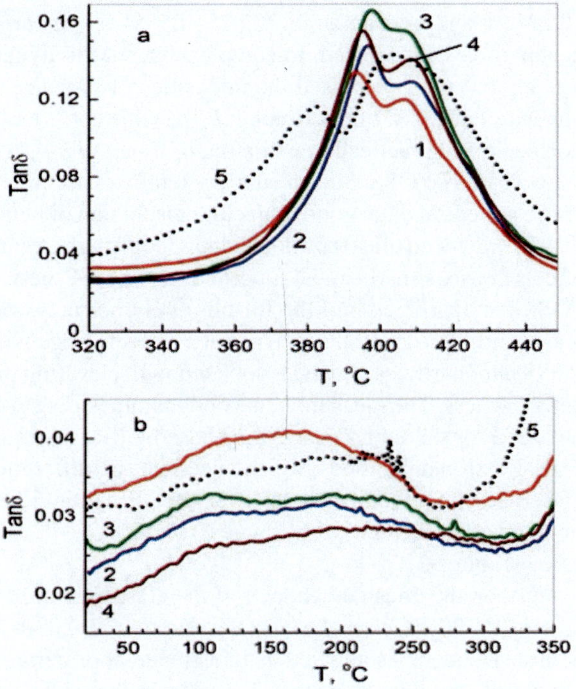

Fig. 2.17 DMA (1 Hz): tanδ (T) relaxation spectra obtained in the temperature regions of glass transition **a** and *sub*-T_g relaxation **b** for cured unfilled PT-30-based CER matrix (1) and PT-30-based CER/ECH-POSS nanocomposites with 0.025 (2), 0.1 (3), 0.5 (4), and 10 wt. % ECH-POSS nanoparticles (5) (Bershtein et al. 2020)

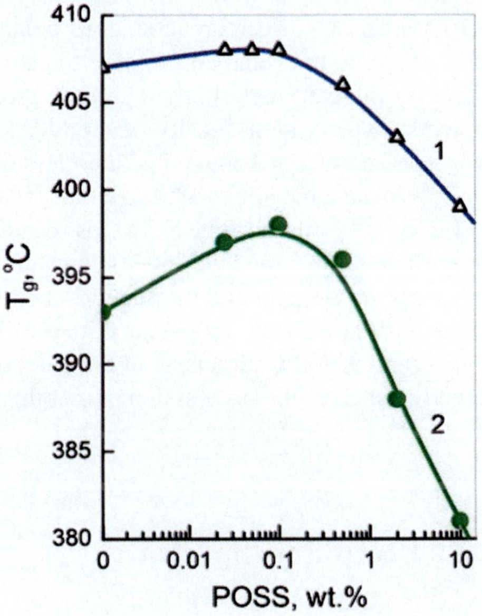

Fig. 2.18 Glass transition temperatures of the PT-30-based CER/ECH-POSS nanocomposites as the functions of ECH-POSS content. T_g values were determined as the T_{max} temperatures of the tanδ (T) doublets shown in Fig. 2.17 (Bershtein et al. 2020)

Ultra-low and low amounts (0.025–0.5 wt.%) of introduced reactive ECH-POSS nanoparticles suppressed, to some extent, matrix dynamics in the sub-T_g temperature region, viz., decreased the intensities of broad relaxation peak. Meantime, the opposite tendency of decreasing T_g s, with considerable broadening glass transition peak, and practically restoration of broad sub-T_g relaxation peak may be seen at 10 wt.% ECH-POSS. Such results are totally expectable being in accordance with the above structural data: good molecular dispersion of nanoparticles in the first case and their pronounced clusterization in the second case. Again, the obvious origin of these effects consists in chemical modification of CER network due to embedding ECH-POSS nanoparticles into the forming polymer network as the additional inorganic nodes and the "constrained dynamics" effect. Negative influence of 10 wt.% ECH-POSS nanoparticles may be associated with elevating the network heterogeneity and defectiveness. The measurements of dynamic (storage) modulus E'(T) dependencies, obtained for the cured neat PT-30-based CER matrix and three PT-30-based CER/ECH-POSS nanocomposites, showed that the difference in their rigidities could be discerned only at temperatures over the curing temperature of 300 °C; the tendency of decreasing E' might be seen at 10 wt.% ECH-POSS (Bershtein et al. 2016) (see below Table 2.2).

Further, the unusual behavior of the glass transition was detected by DMA in the case of PT-30-based materials (Bershtein et al. 2020) The tanδ (T) plots measured at three frequencies allowed us to estimate approximately the "apparent" activation energy of the glass transition E_{act}. The value of T_g was virtually unchanged with variation in frequency (within an accuracy of ± 2 °C), which corresponded to E_{act} > 1000 kJ mol^{-1}. Such an anomalous behavior of the relaxation glass transition is very similar to the behavior of phase transitions. One could presumably explain this effect by the extremely high degree of cooperativity of this transition in polymer networks with a high density of cross-linking. It should be noted that we found the same anomalous behavior of the glass transition in the ultra-high-heat-resistant phthalonitrile nanocomposites (see Part II).

Figure 2.19 demonstrates how post-curing of the PT-30-based nanocomposite at an optimal temperature of 360 °C leads to two simultaneous effects—in the DSC curve and in the dynamic mechanical spectrum. It can be seen that in this analysis of the post-cured nanocomposite, both the exotherm of post-polymerization at 400–450 °C and the relaxation peak of the glass transition around 400 °C disappear. This directly indicates the suppression of mobility in this nanocomposite due to additional polymerization of the matrix network.

2.4 Creep Rate Spectra

The interesting data have been obtained also by laser-interferometric creep rate spectroscopy (CRS). This original high-resolution technique has been developed in our Materials Dynamics Laboratory at Ioffe Institute (Bershtein and Yakushev 2010). This review article contains detailed information about CRS experimental setups,

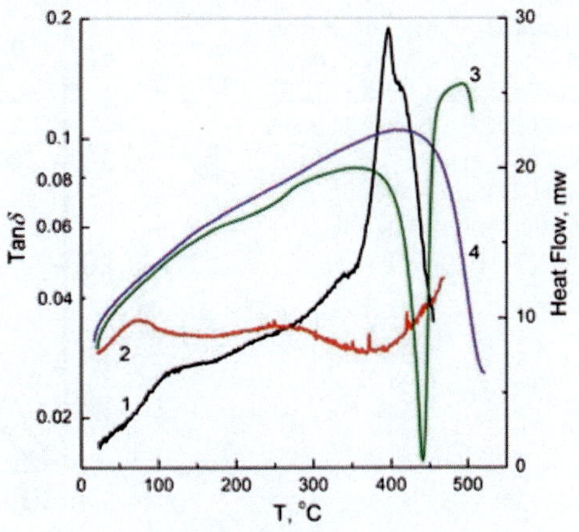

Fig. 2.19 DMA (tanδ) (1, 2) and DSC (3, 4) data obtained for the PT-30-based CER/ECH-POSS (0.05 wt.%) nanocomposite cured by heating from 20 to 300 °C (1, 3) and post-cured for 0.5 h at 360 °C (2, 4) (Bershtein et al. 2020)

working methodology, and numerous examples of the application of this method in physical research. CRS, being a highly sensitive discrete method of relaxation spectrometry and thermal analysis, has contributed to solving various problems in polymer physics and materials science. Thus, this method has been successfully applied to the problems of molecular dynamics analysis, micro- and submicroplasticity, evaluation of dynamic heterogeneity and its relationship with the structure of materials; studies of creep kinetics both at a given temperature and in a wide temperature range, etc.

CRS consists in precisely measuring creep rates at a constant low mechanical stress as a function of temperature by means of a laser interferometer based on the Doppler effect. The stress values are chosen as capable of inducing sufficient creep rates to be measured while maintaining also good spectral resolution and preventing a premature rupture of a sample.

Figure 2.20 **a** shows the creep rate spectra obtained at a tensile stress of 0.4 MPa over the temperature range between 180 and 340 °C for the neat DCBE-based CER matrix and CER/ECH-POSS nanocomposites with 0.1 or 0.5 wt.% nanofillers. Several points should be noted in this case.

In general, the "constrained dynamics" effect is clearly observed in the nanocomposites, which is more pronounced in the sample with ultra-low nanoparticle content (0.1 wt.%). It manifests itself in the displacement of the onset of spectra by 10° to higher temperatures and enhancing high-temperature creep resistance. Secondly, we observe the discontinuous (discrete) nature of these spectra, including several overlapping peaks around T_g; this indicates dynamic heterogeneity in the temperature regions of 200–260 °C in the case of neat CER matrix and 210–320 °C for the nanocomposites. Finally, the temperature of the sharp acceleration of the creep process in nanocomposites increases significantly, and the failure of the neat matrix occurs at 270° and of the nanocomposite—at 320 °C.

The positive influence of adding POSS nanoparticles may be seen also in the case of using another, DCBA-based CER matrix (Fig. 2.20).

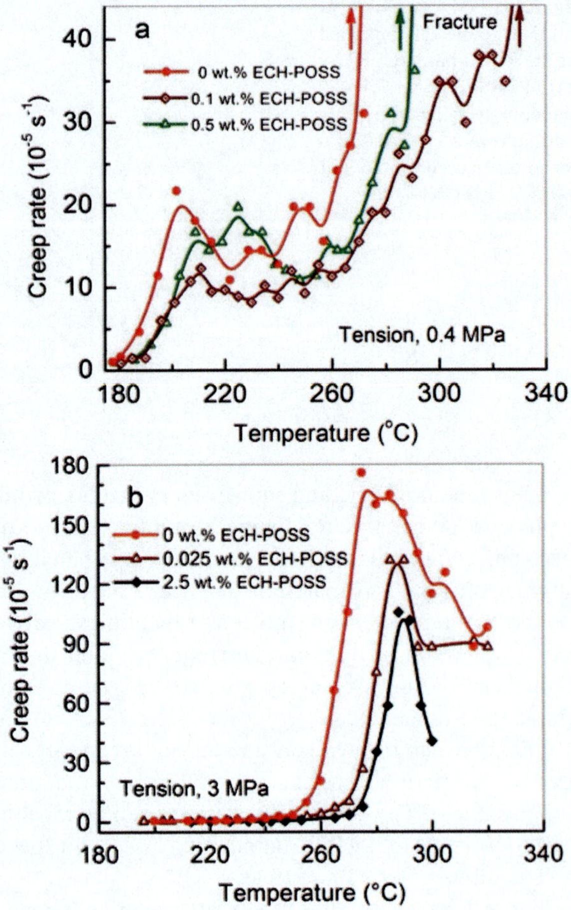

Fig. 2.20 Creep rate spectra obtained at the indicated tensile stresses for **a** DCBE- and **b** DCBA-based matrices and the ECH-POSS-containing nanocomposites based thereon (Bershtein et al. 2016)

Figure 2.21 shows the creep rate spectra obtained at a tensile stress of 0.4 MPa for the neat PT-30-based CER matrix and the PT-30-based CER/ECH-POSS (0.05%) nanocomposite. One can see that the extremely low level of creep rates up to ~ 250 °C with the peak having T_{max} at ~ 260 °C are observed under these conditions for the neat matrix.

At the same time, introducing ultra-low content of ECH-POSS nanoparticles results in increased creep resistance: a very low level of creep rates extends up to 290 °C, and the main peak decreases with shifting T_{max} up to ~ 320 °C. To compare, the inset is also presented in Fig. 2.21 showing the creep rate spectra of the CER matrix and its nanocomposite with 0.1% ECH-POSS obtained by polymerization of DCBE monomer. In this case, the positive influence of nanoparticles is observed as well but the higher creep rates are observed starting already from 180 to 200 °C. Thus, the advantage of materials obtained from PT-30 oligomer regarding creep resistance is obvious.

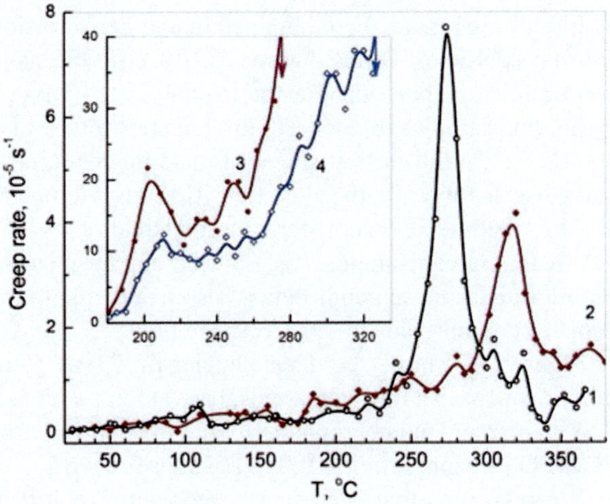

Fig. 2.21 Creep rate spectra were obtained at a tensile stress of 0.4 MPa for the unfilled PT-30-based CER matrix (1) and the PT-30-based CER/ ECH-POSS (0.05%) nanocomposite (2). To compare, the inset shows the creep rate spectra obtained at the same stress for the cured DCBE-based CER matrix (3) and its nanocomposite with 0.1 wt.% ECH-POSS (4) (Bershtein et al. 2020)

2.5 Thermal Stability

TGA study of the thermo-oxidative decomposition of DCBE-based CER network and CER/ECH-POSS nanocomposites (Bershtein et al. 2015) showed that all the studied samples showed no appreciable mass loss when heated to 400 °C; the mass loss of about 2% was due to the removal of absorbed water. The TGA curves obtained at heating rate of 20 °C/min indicated two stages of intense mass loss: in the region of 400–470 °C (stage I) and at 500 to 700 °C (stage II). Stage I was associated with the degradation of the skeleton of the CER network including the triazine cycles. One could assume that the resulting free radicals and reactive atomic groups interacted with oxygen to form more thermally stable structures that were destroyed at higher temperatures in stage II.

As shown (Bershtein et al. 2015), the resistance to thermo-oxidative degradation did not change for nanocomposites with low POSS content; moreover, increasing the POSS content up to 10 wt.% resulted in a significant decrease in the resistance to thermo-oxidative degradation at stage 1 of the process. The fact is that with increasing content of ECH-POSS nanoparticles in the matrix structure becomes more linear fragments (oxazolidinone connections), i.e., the degree of cross-linking of the CER network becomes lower. Linear fragments of the network degrade easier than perfect heterocyclic networks with a high concentration of rigid cross-links.

It was assumed that the interaction of free radicals and reactive atomic groups formed in stage 1 of the composite degradation process with ECH-POSS nanoparticles generated some hybrid structures with increased thermal stability (SiO_xC_y type, as suggested by Song et al. (Song et al. 2008)); this allows the degradation process to "shift" to higher temperatures. However, it should still be recognized that, in general, the addition of POSS did not noticeably increase the resistance of the material to the thermo-oxidative degradation at temperatures above 400 °C.

Interesting results were obtained in another series of experiments to evaluate the thermal stability of DCBE-based CER/ECH-POSS nanocomposites, when using the non-traditional approach (Bershtein et al. 2015) These experiments were performed by holding samples in an inert nitrogen atmosphere at different temperatures below 400 °C, i.e., at the earliest stages of material decomposition, under conditions of no mass loss. In this case, the glass transition onset temperature, T_g', was determined by the DSC method after short-term thermal treatments (10 min) under inert atmosphere in a scanning calorimeter. It should be emphasized that the evaluation of thermal stability under these conditions is also of practical interest when using these high-temperature materials in the aerospace field.

Figure 2.22 shows the dependencies of T_g' as a function of temperature of 10-min treatment in a nitrogen atmosphere at temperatures from 300 to 380 °C for neat CER matrix and nanocomposites based thereon and containing ultra-low amounts of ECH-POSS nanoparticles, 0.01, 0.025, or 0.1 wt.%.

It can be seen that even such a short-term treatment of the neat matrix at 320–360 °C already leads to an essential decrease in the value of T_g' caused by the destructive process that has started.

At the same time, ultra-low POSS additives, starting from 0.01 wt.%, exhibited a positive influence on the thermal stability of the CER network. The maximal positive effect was observed at 0.025 wt.% POSS when after the treatment at 380 °C glass transition started at a higher temperature than for neat CER matrix before thermal treatment.

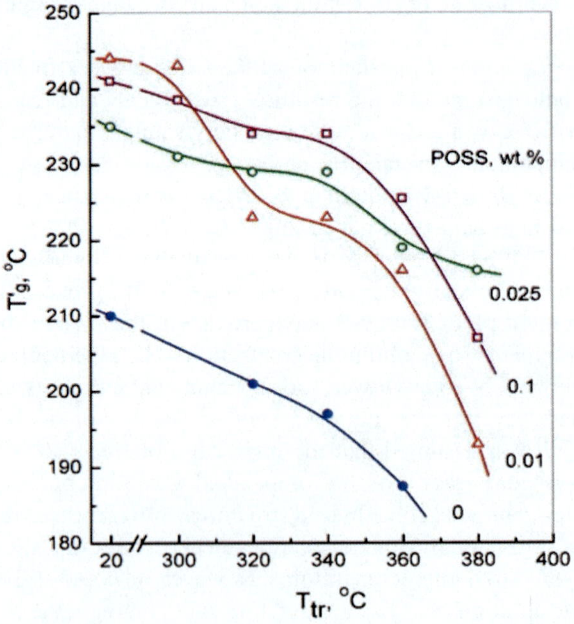

Fig. 2.22 Thermal degradation onset as estimated by DSC measuring temperatures of glass transition onset for the DCBE-based CER/ECH-POSS nanocomposites after their 10-min treatment at different temperatures T_{tr} in a nitrogen atmosphere. Nanofiller contents are indicated (Bershtein et al. 2015)

Zhang et al. (2014) investigated a series of CER/ECH-POSS nanocomposites with high nanoparticle content, 5, 10, 15, and 20 wt.%. This work showed, in particular, an increase in the impact and flexural strengths, and the dynamic modulus E' in the rubbery state in the nanocomposites. TGA also showed that these nanocomposites have improved thermal stability. Thus, the temperature of the beginning of decomposition with mass loss (Ti) of the CER/ECH-POSS (10 wt.%) nanocomposite was 426 °C, by 44° higher than for the neat matrix (Zhang et al. 2014).

Figure 2.23 presents the TGA data obtained at temperatures between 20 and 700 °C in nitrogen or air atmospheres for the PT-30-based CER matrix and for the PT-30-based CER/ECH-POSS (0.5%) nanocomposite. The data characterize thermal stability and thermo-oxidative decomposition, respectively. Additionally, to compare, the TGA curve for the CER/ECH-POSS (0.5 wt.%) nanocomposite, where DCBE monomer was polymerized, is also represented. It is clear that these materials are completely stable under these conditions up to about 400 °C, regardless of the nature of the environment (the insignificant loss of mass is due to the removal of absorbed water). The first stage of degradation starts at 420 to 430 °C and the process continues up to about 550 °C also regardless of the nature of the environment and ECH-POSS additive (Fig. 2.23, curves 1–4). At this stage, about 20% mass is lost for the PT-30 based nanocomposite but up to about 40% for the DCBE-based nanocomposite (Fig. 2.23, curve 5).

Further, the temperature T ≈ 550 °C manifests itself as a certain “bifurcation point”: it is from this temperature that the TGA curves obtained in the measurements in the nitrogen and air mediums sharply diverge due to the “switching on” of intensive thermo-oxidative reactions into the degradation process in the air medium. As a result, curves 3 and 4 go abruptly down, whereas the additional destructive processes are insignificant in nitrogen atmosphere up to 700 °C (curves 1 and 2 in Fig. 2.23). Char residues at 700 °C for the neat PT-30-based matrix were equal to about 68 wt.% and 25 wt.% in these two mediums, respectively. The covalent “embedding” of ECH-POSS nanoparticles into this matrix resulted in a certain increase in thermo-oxidative

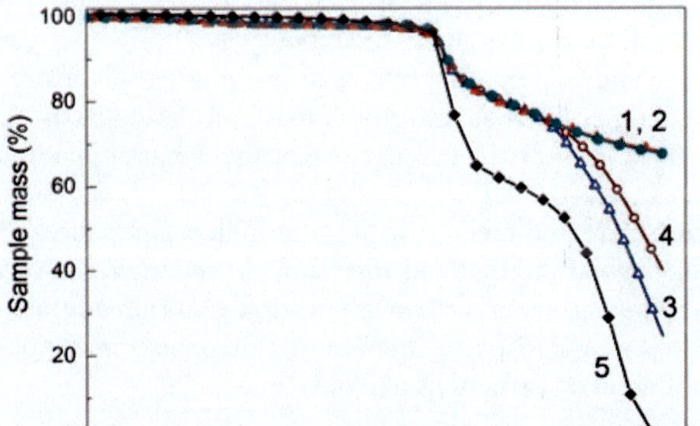

Fig. 2.23 TGA curves obtained at heating with the rate of 20 °C/min in nitrogen (1, 2) and air (3, 4, 5) mediums for the unfilled PT-30-based CER matrix (1, 3) and the CER/ECH-POSS (0.5 wt.%) nanocomposite (2, 4). To compare, curve (5) is given for the CER/ ECH-POSS (0.5 wt.%) nanocomposite synthesized from DCBE monomer (Bershtein et al. 2020)

stability in the air: the char residue at 700 °C increased from 25 wt.% to 40 wt.%. Meanwhile, in the case of nanocomposite synthesized on the basis of DCBE monomer char residue was close to zero at 700 °C when heated under the same conditions in the air environment. At the same time, for the nanocomposite synthesized from DCBE monomer, char residue was close to zero at 700 °C under air conditions (Fig. 2.23, curve 5). Enhancing thermo-oxidative stability in the PT-30-based CER/ECH-POSS (0.5 wt.%) nanocomposite as compared to that for the unfilled matrix might be determined, presumably, with the changes in degradation kinetics due to the barrier effect, some suppression of dynamics in the matrix, restricting oxygen diffusion, etc.

Lin and Song (2018) showed that the thermal degradation process of the CER network in the PT-30-based CER/POSS nanocomposites involves three types of reactions: (a) destruction of the bonds between the phenyl and triazine rings; (b) decay of the triazine rings, and (c) destruction of the benzene rings. According to their data, the introduction of POSS led to a significant improvement in the thermal stability of the matrix. Thus, when performing experiments in air, the temperature of complete decomposition of the material increased with the introduction of 1 wt.% POSS by 146 °C, and the activation energy of the degradation process of the composite was significantly higher than in the process of degradation of the neat matrix.

Thus, depending on the content of ECH-POSS nanoparticles in the CER-based nanocomposites, they have an ambivalent impact on the matrix dynamics, nanocomposite properties, and on the nature of the relationship between their nanostructure and properties. The most general and significant result of the considered works is the experimental discovery of the fact of the most pronounced positive effect from the introduction of ultra-low concentrations ($\ll$1 wt.%) of reactive ECH-POSS nanoparticles, covalently "embedded" in the densely cross-linked CER matrix network.

As the POSS nanoparticle content in nanocomposites increased, especially from 2 wt.% to 10 wt.%, when such structural formations as nanorod- or platelet-like ones and POSS-enriched nanodomains appeared in nanocomposites, this led to a reduced, almost zero or even negative influence of the nanoparticles on CER properties.

Another important result of the above studies is the following. Nanocomposites derived from multi-functional PT-30 oligomer can exhibit higher thermal and mechanical properties than nanocomposites synthesized from two-functional DCBE monomer. This superiority consists in particular in the increase of T_g from 240–300 °C to 400 °C; the increase of the dynamic modulus' from 2 to 4 GPa at 20 °C and from 0.1 GPa to 2 GPa at 300 °C; higher thermal stability at 450–700 °C and increased creep resistance at high temperatures. Finally, it is clear that due to their good thermal and mechanical properties the considered nanocomposites are promising materials for application (and already are applied!) in extreme conditions—in aerospace, submarine, microelectronic industries, and in "hot zones" of different constructions and devices.

2.6 Influence of Variation in the Chemical Structure of the Monomer on the Properties of the CER/POSS Nanocomposites

Above, the superiority of the PT-30-based nanocomposites over the DCBE-based nanocomposites has been shown. In (Bershtein et al. 2016), we synthesized the CER/ECH-POSS nanocomposites with four CER matrices of different chemical structures containing nanoparticles in a wide concentration range, from 0.01–0.025 wt.% up to 10 wt.%. This study was aimed at analyzing the effect of the chemical structure of the matrix on molecular structure—composition—T_g—modulus—thermal properties relationships in these materials.

1,1′-bis(4-cyanatophenyl)ethane (dicyanate ester of bisphenol E (DCBE)), under the trade name PRIMASET™ LECy L-10; 2,2′-bis(4- cyanatophenyl)isopropylidene (dicyanate ester of bisphenol A (DCBA)), under the trade name PRIMASET™ BADCY; oligo(3-methylene-1,5-phenylene-cyanate), cyanated oligomer under the trade name PRIMASET™ PT-30, and 2,2′-bis(4-cyanatophenyl)-1,1,1,3,3,3-hexafluoroisopropylidene (dicyanate ester of hexafluorobisphenol A, 6F-DCBA) under the trade name AroCy F-10, were used as initial products. Their formulas are presented in Table 2.1. Curing neat matrices and nanocomposites has been performed over the temperature range from 25 to 300 °C at the heating rate of 0.5 °C/min. The combined FTIR and far-IR spectroscopy, DMA, DSC, and laser-interferometric CRS have been used for their analysis (Bershtein et al. 2016).

Again, as before, far-IR spectra were measured in the range of 20 to 280 cm^{-1} for both neat matrices and nanocomposites based thereon containing ultra-low (≪1 wt.%) or elevated (1, 2.5 or 10 wt.%) amounts of ECH-POSS nanoparticles. In these

Table 2.1 Chemical structures of monomers for CER matrices studied (Bershtein et al. 2016)

Name	Chemical structure
Dicyanate ester of bisphenol E, DCBE	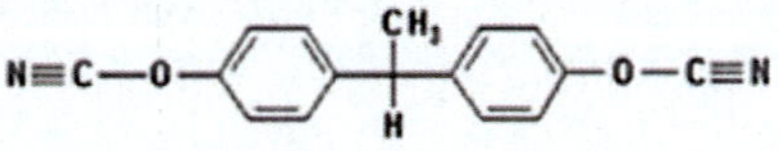
Dicyanate ester of bisphenol A, DCBA	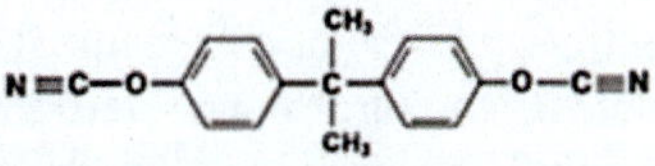
Dicyanate ester of hexafluorobisphenol A, 6F-DCBA	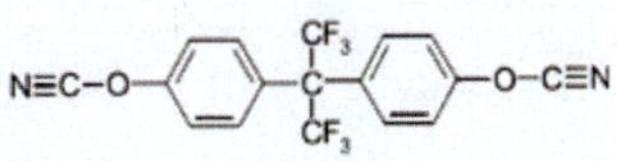
Oligo (3-methylene-1,5-phenylene-cyanate), PT-30	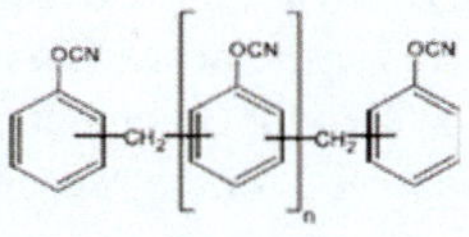

Table 2.2 Dynamic (storage) modulus of the four types of CER/ECH-POSS nanocomposites studied at different temperatures below 400 °C (Bershtein et al. 2016)

CER matrix prepared from	ECH-POSS [wt%]	Dynamic modulus E′ (GPa, at 1 Hz) at different temperatures (°C)						
		20	200	250	300	340	360	380
DCBE	0	1.8	1.3	0.1	≪ 0.1	–	–	–
	0.025	1.7	1.4	0.5	≪ 0.1	–	–	–
	0.05	1.5	1.2	0.3	≪ 0.1	–	–	–
	0.5	2.4	1.8	0.3	≪ 0.1	–	–	–
DCBA	0	3.0	2.5	1.2	< 0.001	–	–	–
	0.025	2.7	2.1	1.8	< 0.001	–	–	–
	2.5	3.1	2.6	2.1	< 0.001	–	–	–
6F-DCBA	0	2.7	2.1	1.8	~ 0.001	–	–	–
	0.025	2.8	2.3	1.9	~ 0.001	–	–	–
	0.1	2.6	2.0	1.6	~ 0.001	–	–	–
	1	3.0	2.5	2.0	~ 0.001	–	–	–
PT-30	0	3.6	2.5	2.1	1.6	1.2	1.0	0.8
	0.025	3.7	2.4	2.0	1.6	1.3	1.1	0.9
	0.5	3.8	2.4	2.0	1.6	1.3	1.1	0.9
	10	3.5	2.3	2.0	1.6	1.0	0.7	0.4

experiments, it was found that the covalent "embedding" of ECH-POSS nanoparticles into CER networks leads to different "constrained dynamics effects" in the vibrational dynamics of the matrices, depending both on their chemical structure and on the nanoparticle content. The most significant changes in the matrix dynamics were observed in the presence of ultra-low contents of ECH-POSS units (0.025–0.1 wt.%); this was caused not only by their chemical grafting to the matrix network but also by the good dispersity of nanoparticles at the molecular level (prevention of their clustering process). And again, as in the above experiments, at high loading of the matrix with ECH-POSS particles (10 wt.%), there were no noticeable effects of the suppression of vibrational dynamics of the matrix. Thus, the observed changes in the far-IR spectra were in good agreement with the structural data, e.g., with the highly pronounced structural nanoheterogeneity of the composites with 10 wt.% ECH-POSS. It should also be noted that the effect of nanoparticles on the vibrational dynamics of the polymer matrix decreased gradually in the row of matrices obtained from the following products: DCBE > DCBA > 6F-DCBA > PT-30 (Bershtein et al. 2016).

Further, the results of DMA as tanδ and loss modulus E'' versus temperature dependencies measured at 1 Hz for the neat CER matrices and the nanocomposites based thereon, with different ECH-POSS contents, were obtained. Significantly different behavior was already observed for the four neat CER matrices. Thus, at a frequency of 1 Hz, tanδ maxima temperatures (temperature positions of the main glass transition peak) were equal to 248°, 275°, 300°, and about 400 °C for the CER matrices prepared from DCBE, DCBA, 6F-DCBA, and PT-30, respectively. The introduction of ECH-POSS nanoparticles into the CER matrix resulted in the substantial displacement of glass transition peak in the cases of DCBE- and DCBA-based matrices but influenced relatively slightly in the cases of networks prepared from 6F-DCBA and PT-30. The effects observed depended, of course, on the ECH-POSS content in composites.

Figure 2.24 shows the temperatures of glass transition onset, T_g', completion, T_g'', and T_g s as the functions of ECH-POSS content in the nanocomposites with different matrices. The measurements were performed by DMA and DSC. Depending on the chemical structure of the matrix, one can observe quite different influence of the introduced POSS nanoparticles on the indicated temperatures in the glass transition of nanocomposites.

The maximum positive effect is observed in the case of DCBE-based matrix, and it is already achieved with the introduction of only 0.025–0.1 wt.% ECH-POSS nanoparticles. In this case, T_g', T_g, and T_g'' increased by 50°, 30°, and 20 °C, respectively, i.e., the narrowing of the glass transition range occurred. The effects decreased after incorporating 0.5–2.0 wt.% ECH-POSS into this matrix, and at 10 wt.% ECH-POSS, there was again a strong broadening of the glass transition, and the positive effect of nanoparticles practically disappeared. These results as indicated above were due to structural reasons: good molecular dispersity at 0.025 wt.% ECH-POSS and

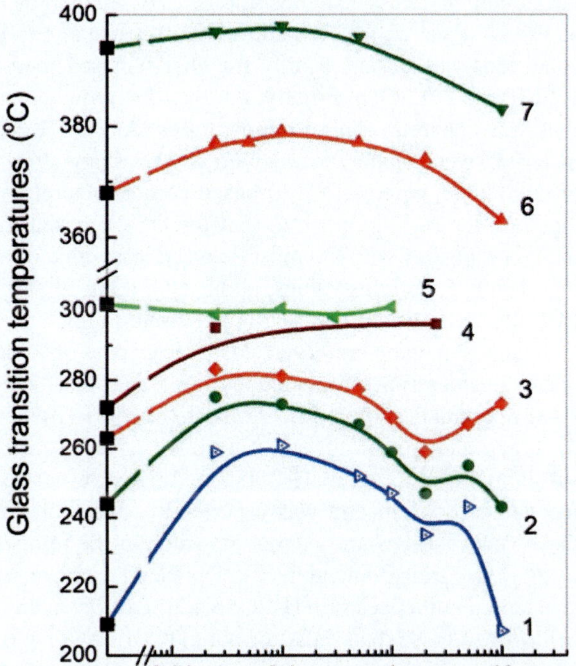

Fig. 2.24 Glass transition temperatures of the CER/ECH-POSS nanocomposites as a function of ECH-POSS content: 1, 2, 3—T_g', T_g, and T_g'', respectively, for the DCBE-based nanocomposites (DSC); 4—T_g for the DCBA-based nanocomposites (DMA); 5—T_g for the 6F-DCBA-based nanocomposites (DMA); 6, 7—T_g for PT-30-based composites (DSC and DMA, respectively). Black squares relate to neat matrices (Bershtein et al. 2016)

increased structural heterogeneity with the formation of ECH-POSS aggregates at 10 wt.% ECH-POSS.

For the composites with the highest cross-link density and T_g close to 400 °C, obtained from PT-30 oligomer, qualitatively the same dependencies were observed as for DCBE-based materials in the both DSC and DMA experiments, but the effects of T_g increasing in the nanocomposites were of a few degrees only (curves 6 and 7). Another dependencies were observed for the nanocomposites with DCBA- and 6F-DCBA- based matrices (Fig. 2.24, curves 4 and 5), where for the former matrix T_g increased by about 20 °C after introducing 0.025 wt.% or 2.5 wt.% ECH-POSS, whereas for the latter matrix $T_g \approx 300$ °C for all the nanocomposites and neat matrix, i.e., there was no the influence of nanoparticles on the T_g value.

The values of dynamic (storage) module E' of the materials under study, estimated by DMA at 1 Hz over the temperature range from 20 to 380 °C, are presented in Table 2.2. A few points are noteworthy herein. First, the modulus E' of the studied matrices at 20 °C increases in a row: DCBE < 6F-DCBA < DCBA < PT-30 from 1.8 GPa to 3.6 GPa. Secondly, ECH-POSS additives only slightly changed rigidity of the matrices at temperatures from 20 to 200 °C but the substantial and very different changes in rigidity of matrices and nanocomposites based thereon manifested themselves at high temperatures starting from about 250 °C. At this temperature, very strong dropping of E' value was observed only for DCBE-based matrix, although introducing 0.025 wt.% ECH-POSS resulted in a partial retaining rigidity. Meantime, three other matrices and nanocomposites based thereon retained rigidity at the E' level of 1 to 2 GPa at 250 °C. The positive influence of ECH-POSS additives under these conditions was observed only for DCBA-based matrix (increasing modulus from 1.2 MPa to 1.9 MPa). Finally, the obvious great superiority of PT-30-based materials was apparent at high temperatures. At 300 °C, DCBA- and 6F-DCBA-based materials were characterized with a very sharp drop of rigidity, by a few orders of magnitude, whereas PT-30-based composites retained E' value of 1.6 MPa, i.e., they had the rigidity close to that for DCBE-based materials at 20 °C. Moreover, these composites with the most densely cross-linked matrix network retained rather high level of modulus values, about 1 GPa, up to 380 °C although the influence of ECH-POSS additives was negligibly small in this case.

In addition to the improved elastic properties of some of the CER nanocomposites under consideration at high temperatures, their resistance to high-temperature creep is also important in some applications. In this work (Bershtein et al. 2016), we measured the creep rate spectra for the materials derived from DCBE and DCBA, where the substantial impact of ECH-POSS additives on network vibrational dynamics and mechanical loss spectra was described above. The creep rate spectra obtained for these neat matrices and nanocomposites in the temperature region within or close to the glass transition range were presented above in Fig. 2.20. One could see a substantial influence of ECH-POSS additives (even their ultra-low amounts) on creep behavior of these materials: for both DCBE- and DCBA-based nanocomposites the temperature of the sharp creep acceleration increased by about 20–30 °C regarding that for neat matrices.

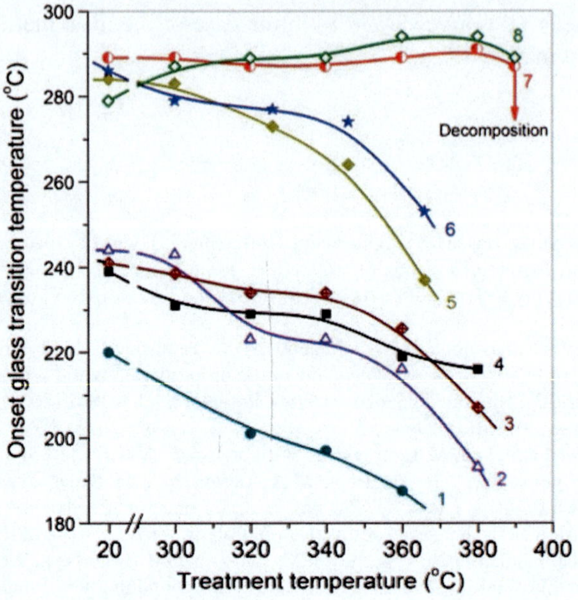

Fig. 2.25 Thermal degradation onset as estimated by DSC measuring temperatures of glass transition onset for three different CER matrices and CER/ECH-POSS nanocomposites after their 10-min treatment at different temperatures T_{tr} in nitrogen atmosphere: 1—DCBE-based matrix, 2—ditto + 0.01 wt.% ECH-POSS, 3—ditto + 0.1 wt.% ECH-POSS, 4—ditto + 0.025 wt.% ECH-POSS, 5—DCBA-based matrix, 6—DCBA-based matrix + 0.025 wt.% ECH-POSS, 7—6F-DCBA-based matrix, and 8—6F-DCBA-based matrix + 0.1 wt.% ECH-POSS (Bershtein et al. 2016)

At last, Fig. 2.25 shows the results of studying thermal degradation onset, as estimated by DSC measuring temperatures of glass transition onset, T_g', after their 10-min treatment at different temperatures $T_{tr} = 300$–380 °C in nitrogen atmosphere, for three CER matrices and their nanocomposites. This is actually an estimate of thermal stability, when destructive processes are still going on without loss of mass.

One can see that the thermal stability of the CER matrices under study increased in a row: DCBE < DCBA < 6F-DCBA. Even a short-term thermal treatment of neat DCBE- or DCBA-based matrices in inert atmosphere at temperatures of 320° to 360 °C resulted in an essential decrease of T_g' values because of the onset of destructive processes in the networks (curves 1 and 5). Contrarily, 6F-DCBA-based matrix network remained undamaged up to 380 °C; it decomposed at 390 °C (curve 7). Figure 2.25 shows that ultra-low amounts of ECH-POSS additives, starting from 0.01 wt.%, already influenced the short-term thermal stability of CER networks at temperatures below 400 °C. It is obvious most distinctly for DCBE- based composites (curves 2–4) where the maximal positive effect was observed at 0.025 wt.% ECH-POSS: after the treatment at 380 °C glass transition started at the same high temperature as for neat CER matrix before thermal treatment. Nanoparticles exerted less effect on the thermal stability of the DCBA-based matrix (curves 5 and 6) and

had no influence at all (curves 7 and 8) in the case of the most thermally resistant 6F-DCBA-based matrix.

References

L. Baikova, T. Pesina, D. Sakseyev, A. Fainleib, V. Bershtein, Hybrid cyanate ester resin-based nanocomposites: increased indentation size effect due to anomalous composition of micron subsurface layer. Polym. Test. **53**, 15–18 (2016). https://doi.org/10.1016/j.polymertesting.2016.05.001

V.A. Bershtein, L.V. Karabanova, T.E. Sukhanova, P.N. Yakushev, L.M. Egorova, E. Lutsyk, A.V. Svyatyna, M.E. Vylegzhanina, Peculand elastic properties of hybrid semi-interpenetrating polymer network–3-D diamond nanocomposites. Polymer **49**, 836–842 (2008). https://doi.org/10.1016/j.polymer.2008.01.002

V.A. Bershtein, A.M. Fainleib, L.M. Egorova, O.P. Grigoryeva, D.A. Kirilenko, S.G. Konnikov, V.A. Ryzhov, O.M. Starostenko, P.N. Yakushev, M.A. Yagovkina, J.M. Saiter, The impact of ultra-low amounts of introduced reactive POSS nanoparticles on structure, dynamics and properties of densely crosslinked cyanate ester resins. Europ. Polym. J. **67**, 128–142 (2015)

V.A. Bershtein, A.M. Fainleib, P.N. Yakushev, L.M. Egorova, O.P. Grigoryeva, V.A. Ryzhov, O.M. Starostenko, Thermostable cyanate ester resins and POSS-containing nanocomposites: influence of matrix chemical structure on their properties. Polym. Adv. Technol. **27**, 339–349 (2016). https://doi.org/10.1002/pat.3645

V.A. Bershtein, V.A. Ryzhov, Far-infrared spectroscopy of polymers, in *Advances in Polymer Science*, vol. 114 (Springer, Berlin, Heidelberg, 1994), pp. 43–121. https://doi.org/10.1007/BFb0008694

V.A. Bershtein, P.N. Yakushev, Laser-interferometric creep rate spectroscopy, in *Advances in Polymer Science*, vol. 230, ed. by K. Dusek (Springer, Heidelberg, Dordrecht, London, New York, 2010), pp. 73–219. https://doi.org/10.1007/978-3-642-13532-3

V.A. Bershtein, A.M. Fainleib, P.N. Yakushev, D.A. Kirilenko, L.M. Egorova, O.P. Grigoryeva, V.A. Ryzhov, O.M. Starostenko, High-performance multi-functional cyanate ester oligomer-based network and epoxy-POSS containing nanocomposites: structure, dynamics, and properties. Polym. Compos. **41**, 1900–1912 (2020). https://doi.org/10.1002/pc.25506

E.P. Giannelis, R. Krishnamoorti, E. Manias, Polymer-silicate nanocomposites: model systems for confined polymers and polymer brushes, in *Advances in Polymer Science*, vol. 138 (Springer, Berlin, Heidelberg, New York, 1999), p. 107. https://doi.org/10.1007/3-540-69711-X_3

R. Hosemann, S.N. Bagchi, Direct analysis of diffraction. Acta Cryst **15**, 922 (1962)

K. Liang, G. Li, H. Toghiani, J.H. Koo, C.U. Pittman Jr., Cyanate ester/polyhedral oligomeric silsesquioxane (POSS) nanocomposites: synthesis and characterization. Chem. Mater. **18**, 301–312 (2006). https://doi.org/10.1021/cm051582s

Y. Lin, M. Song, Effect of polyhedral oligomeric silsesquioxane nanoparticles on thermal decomposition of cyanate ester resin. React. Funct. Polym. **129**, 58–63 (2018). https://doi.org/10.1016/j.reactfunctpolym.2017.07.007

L. Song, Q. He, Y. Hu, H. Chen, L. Liu, Study on thermal degradation and combustion behaviours of PC/POSS hybrids. Polym. Degr. Stab. **93**, 627–639 (2008). https://doi.org/10.1016/j.polymdegradstab.2008.01.014

O.M. Starostenko, O.P. Grigoryeva, A.M. Fainleib, J.-M. Saiter, B. Youssef, D. Grande, Effect of epoxy-functionalized POSS on thermal stability of nanocomposites based on crosslinked polycyanurates. Polym. J. (kiev) **36**, 233–244 (2014)

O.M. Starostenko, V.A. Bershtein, A.M. Fainleib, L.M. Egorova, O.P. Grigoryeva, A.B. Sinani, P.N. Yakushev, Thermostable polycyanurate-polyhedral oligomeric silsesquioxane (POSS) hybrid networks: synthesis, dynamics and thermal behavior, in *Molecular Mobility and Order in*

Polymer Systems, ed. by K. Severing, S. Spiegel, Macromolecular Symposia, vol. 316 (John, Ltd, 2012), pp. 90–96. https://doi.org/10.1002/masy.201290018

Z. Zhang, G. Liang, X. Wang, Epoxy-functionalized polyhedral oligomeric silsesquioxane/cyanate ester resin organic-inorganic hybrids with enhanced mechanical and thermal properties. Polym. Intern. **63**, 552–559 (2014). https://doi.org/10.1002/pi.4557

Chapter 3
CER/Montmorillonite Nanocomposites

Abstract The results of physical studies of nanostructure, dynamics, mechanical and thermal properties of DCBE-based CER/amino-MMT nanocomposites are presented. The 2D MMT nanolayers content in these composites varied from 0.01 wt.% to 5 wt.%, and their analysis was performed by TEM, EDXS, mid-IR, far-IR, XRD, DSC, DMA, CRS, and TGA methods. The process of MMT exfoliation increased with decreasing MMT content in the composite, down to the formation of individual nanolayers at 0.1 wt.%. The ambivalent influence of MMT nanolayers on matrix dynamics, T_g, and thermal stability is shown, with the most positive effect at small and ultra-low MMT contents (0.025–0.5 wt.%). Anomalous composition and mechanical properties of the subsurface micron layers in the CER-based nanocomposites were detected.

3.1 About 2D MMT Nanofiller

Among two-dimensional (2D) nanofillers the leading position is occupied by silicate montmorillonite (MMT) clay nanolayers (Bergaya and Lagaly 2013; Nicolais et al. 2012). This is due to such factors as their high aspect ratio, huge specific surface area, high rigidity and thermal stability of these natural materials, as well as their commercial availability and low cost. The introduction of these nanofillers into polymer systems leads to the formation of nanocomposites with significantly improved characteristics (Paul and Robeson 2008). MMT represents a group of minerals with the general formula $M_x(Al_{4-x}Mg_x)Si_8O_{20}(OH)_4$. It consists of self-organized stacks (tactoids) which are divided with certain interlaminar areas (galleries). In general, real MMT-filled polymer systems can exist in three forms: (a) as conventional composites, when sufficiently thick stacks of MMT nanolayers and their agglomerates are distributed in the polymer matrix; (b) as intercalated systems, when the polymer penetrates into interlaminar MMT galleries; and (c) in the limiting case of completely exfoliated MMT nanolayers about 1 nm thick distributed in the polymer matrix. Most researchers have concluded that the greatest effect of improving the properties of the

V. A. Bershtein and P. N. Yakushev, *High-Temperature Polymer Nanocomposites Based on Heterocyclic Networks from Nitrile Monomers*, Springer Series in Materials Science 334, https://doi.org/10.1007/978-3-031-32943-2_3

polymer matrix can be achieved namely in exfoliated MMT/polymer nanocomposites (Paul and Robeson 2008).

It should be noted that the degree of dispersion of MMT nanoparticles in the initial monomer or oligomer and, eventually, in the resulting polymer matrix depends on several conditions of chemical or physical nature. Since the surfaces of virgin layered silicates are hydrophilic, in the case of using usually low-polar or non-polar monomers in the synthesis of nanocomposites certain difficulties can arise. Therefore, to facilitate wetting and penetration of either non-polar or weak-polar chemicals into interplanar areas, MMT chemical modification is used, in particular, hydrophobization. In addition, this procedure is done in such a way when the organic functional groups grafted to the MMT surface are reactive toward the functional groups of the organic monomer or oligomer, or forming polymer network that provides covalent bonding (hybridization) between nanofiller and matrix.

In recent years, at least two articles (Bershtein et al. 2015; Lin et al. 2013) have been published on CER/MMT nanocomposites. Lin et al. (2013) prepared a series of CER/exfoliated MMT nanocomposites with differently modified nanolayers. They revealed accelerating effects on the curing process of CER (the curing temperature decreased by 65 °C) due to the formation of the CER-MMT coordination complex during the curing reaction.

Bershtein et al. (2015) have performed complex research on nanostructure, molecular dynamics, and properties of DCBE-based CER/amino-functionalized MMT nanocomposites with 0.01 wt.% to 5 wt.% nanofillers. MMT was modified by aminopropyltriethoxysilane and octadecylamine. The DCBE/amino-MMT mixture was cured by heating from 150 to 280 °C at the heating rate of 0.5 °C/min. TEM, EDXS, WAXD, DMA, DSC, TGA, Far-IR, and CRS techniques have been used in this research for the characterization of these nanocomposites.

3.2 Structure

Figure 3.1 presents FTIR spectra of neat DCBE-based CER matrix, amino-MMT, and CER/amino-MMT nanocomposites. The disappearance of absorption bands corresponding to cyanate groups of DCBE monomer (at 2237 cm^{-1} and 2266 cm^{-1}) and the appearance of intensive peaks related to triazine cycles in the cross-linked CER matrix (at 1356 cm^{-1} and 1556 cm^{-1}) have been established. Additionally, the hybridization effect (chemical grafting amino-MMT to CER network) in the systems studied was confirmed. This was shown due to the generation of C = NH group of iminocarbamate bonding as well as some redistribution of the peak intensities in the range of 1120–1290 cm^{-1} because of different C–N stretching vibrations in C–NH and C = NH groups. Some transformation of the shoulder at 1645 cm^{-1} for a neat CER network into a peak for CER/amino-MMT nanocomposite samples could be also discerned. The chemical structure of the CER/amino-MMT nanocomposite is presented schematically in Fig. 3.2.

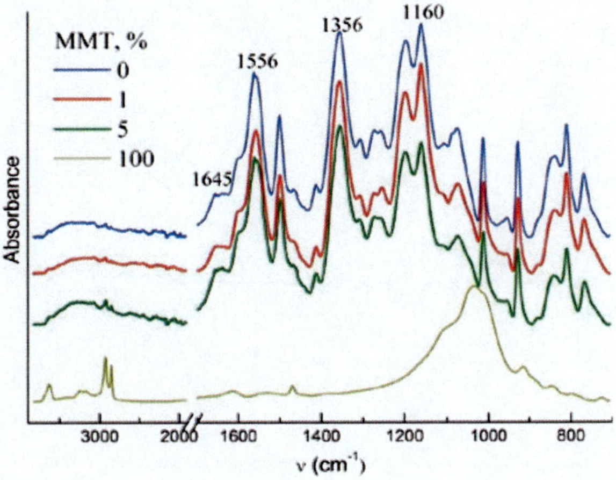

Fig. 3.1 FTIR spectra of neat DCBE-based CER matrix, amino-MMT, and CER/amino-MMT nanocomposites. The spectra are shifted vertically for the sake of clarity (Bershtein et al. 2015)

TEM analysis combined with control EDXS and WAXD measurements allowed us to characterize the morphology of nanocomposites and the character of packing of amino-MMT nanolayers in the matrix. Figure 3.3 displays typical TEM micrographs of the DCBE-based CER/amino-MMT nanocomposites with 0.1, 1, 2, and 5 wt.% nanofiller. No essential structural peculiarities in the CER matrix of nanocomposites in comparison with the structure of the neat matrix were found by TEM. Meanwhile, amino-MMT nanolayers with different degrees of their aggregation into relatively large agglomerates or stacks of different thicknesses, or individual nanolayers could be observed in the nanocomposites with different compositions.

A certain tendency could be followed up, namely, the probability of manifestation of large amino-MMT agglomerates and relatively thick nanolayer stacks decreased with decreasing nanofiller content in the composites, whereas the manifestation of thin amino-MMT stacks and individual nanolayers, i.e., the process of exfoliating nanofiller, became much more pronounced.

Indeed, the nanocomposite with 5 wt.% amino-MMT was characterized mainly both by the absence of essential dispersity of nanolayers and the high degree of non-uniformity of amino-MMT distribution in the matrix (Fig. 3.3a); the typical agglomerates of amino-MMT nanolayers consisted of the nanolayer stacks of about 100 nm in thickness. The nature of these agglomerates was confirmed by the sharply elevated content of Al and Si as it was found in the control EDXS experiments.

Figure 3.4 compares WAXD patterns of mechanical mixture of the powders of amino-MMT particles and polymerized CER (1), and the CER/amino-MMT nanocomposite (2) with the identical contents of amino-MMT (5 wt. %). One can see the essential difference in X-ray diffraction maxima positions and intensities in these patterns. The most intense peak, characterizing amino-MMT packing, shifts from $2\theta = 3.90°$ for the mechanical blend to $2\theta = 2.74°$ for the nanocomposite sample; this shift corresponds to increasing interlayer d-spacing from 2.26 to 3.22 nm.

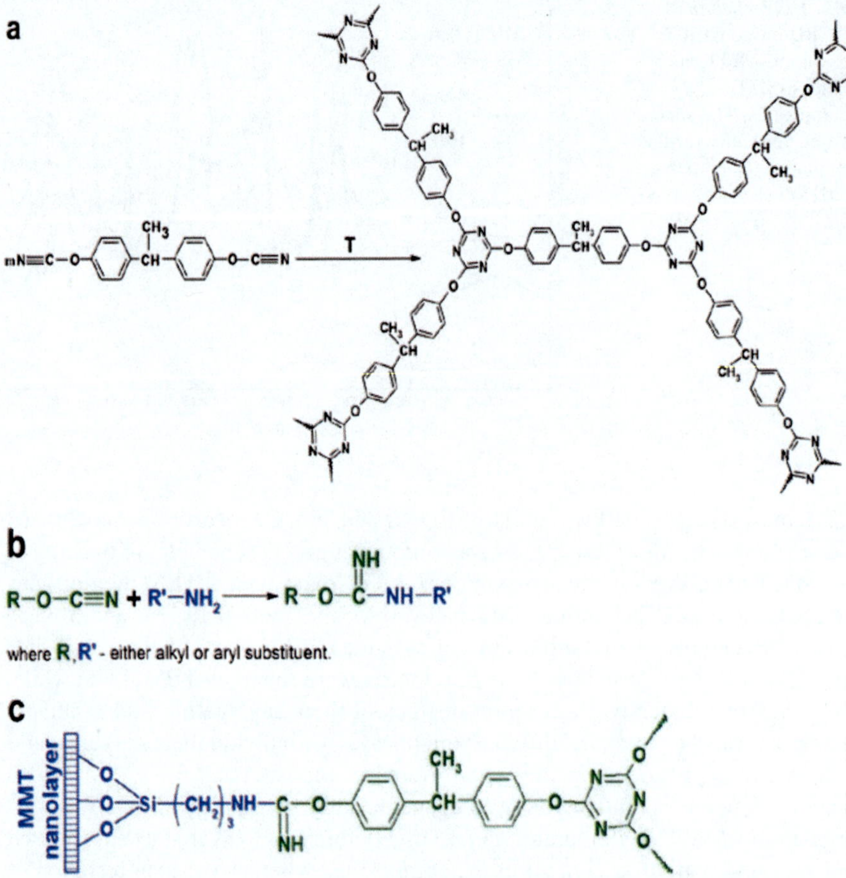

Fig. 3.2 Schemes of synthesis and chemical structure of DCBE-based CER/amino-MMT nanocomposites: polycyclotrimerization of DCBE monomer **a**, chemical interaction between cyanate groups of CER and amino groups of amino-MMT **b**, and the idealized structure of the fragment of DCBE-based CER/amino-MMT nanocomposite **c** (Bershtein et al. 2015)

That indicates slight intercalation of DCBE between nanolayers at the beginning of the process of CER network formation. It should be noted that the typical d-spacing in the natural, non-functionalized MMT mineral is equal to 1.34 nm, i.e., surface amine-containing groups grafted to the MMT nanolayers surface contribute to increasing d-spacing value.

For the nanocomposite with 2 wt. % amino-MMT TEM micrograph changed: the nanofiller aggregates became smaller, and the amino-MMT stacks of about 5–10 nanolayers in thickness could be distinctly discerned (Fig. 3.3b). TEM micrograph changed more substantially for the nanocomposite with 1 wt.% amino-MMT when most typically thin amino-MMT stacks consisting of 3 to 5 nanolayers were observed (Fig. 3.3c and d). At last, at ultra-low amino-MMT contents, e.g., at 0.1% (Fig. 3.3e

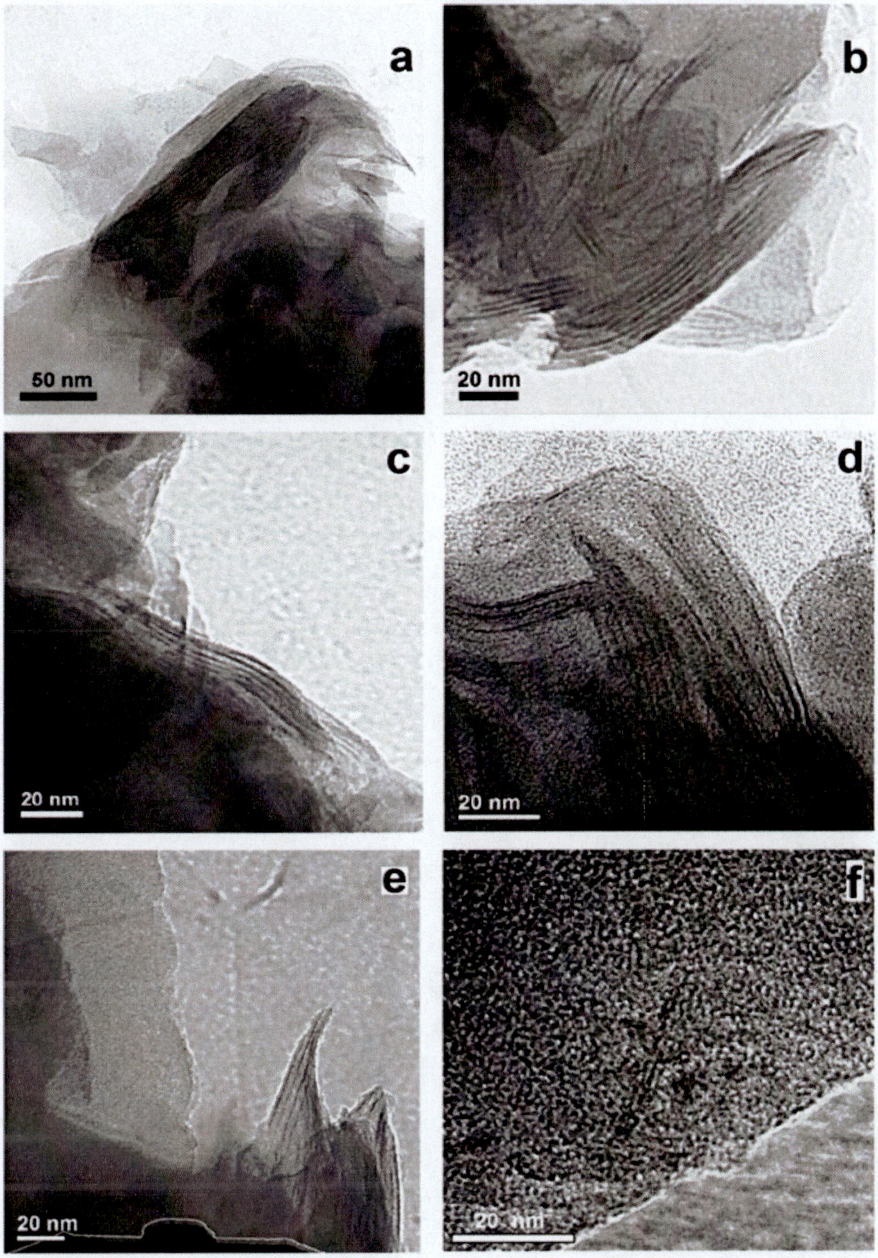

Fig. 3.3 TEM micrographs obtained for the DCBE-based CER/amino-MMT nanocomposites with 5 wt. % **a**, 2 wt. % **b**, 1 wt. % **c, d**, and 0.1 wt. % **e, f** nanoparticles (Bershtein et al. 2015)

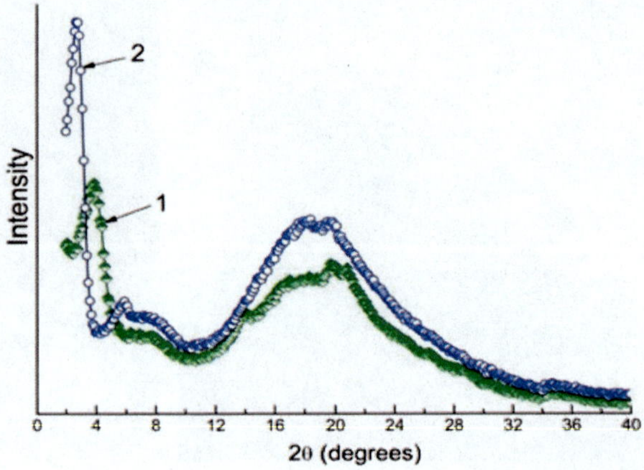

Fig. 3.4 WAXD patterns of DCBE-based CER/amino-MMT (95/5) mechanical mixture (1) and the nanocomposite (2) (Bershtein et al. 2015)

and f) thin, 2 to 3-nanolayer stacks and individual nanolayers (Fig. 3.5) prevailed in the nanocomposite structure. In addition to the TEM micrograph with a single MMT nanolayer of about 1 nm thickness, Fig. 3.5 shows also two EDX spectra taken from 2 nm spots in diameter, just at the individual nanolayer 1 and from the surrounding matrix 2. The attribution of one spectrum to the amino-MMT nanolayer was confirmed by the presence of Al and Si peaks only in this EDX spectrum. The above results suppose that the real interfacial area in the nanocomposites studied may be larger at low and ultra-low amino-MMT contents (if the slight intercalation of the matrix between nanolayers is neglected).

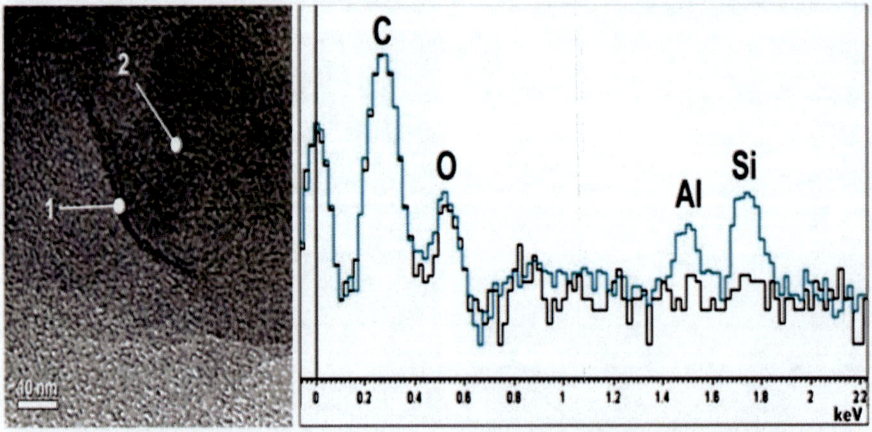

Fig. 3.5 Typical TEM micrograph and the EDX spectra for the DCBE-based CER/amino-MMT (99.9/0.1) nanocomposite. The EDX spectra, taken from 2-nm spots in diameter at the single nanolayer (1) and within the surrounding matrix (2), are also shown (Bershtein et al. 2015)

3.3 Far-infrared Spectra (Matrix Dynamics)

To characterize the changes in molecular (vibrational) dynamics in the CER matrix, caused by covalent incorporating amino-MMT nanoparticles, far-IR spectra of neat matrix and the nanocomposites with 0.05, 0.5, 1, 2, and 5 wt. % nanofiller were measured (Fig. 3.6) (Bershtein et al. 2015). The spectrum of the neat CER matrix displayed the absorption band of small-angle vibrations of triazine and benzene rings with a maximum of 77 cm^{-1}, and the intense band of skeletal torsional vibrations in the CER network with a maximum of 194 cm^{-1}.

Figure 3.6 shows that introducing ultra-low content of amino-MMT nanolayers (0.05 wt. %) exerted a most appreciable "suppressive" impact on the vibrational dynamics of cycles: the band maximum displaced from 77 to 81 cm^{-1}, and the absorption coefficient k decreased from 9.9 cm^{-1} to 9.2 cm^{-1}. The disturbance in the system of skeletal torsional vibrations, with some constraining network dynamics by amino-MMT, was confirmed by decreasing absorption at 160–180 cm^{-1}.

Such changes in molecular dynamics of the matrix were possible, undoubtedly, due to satisfactory (basically as individual layers or 2–3 nanolayer stacks) dispersity of amino-MMT in the matrix, with the formation of multiple CER/nanolayer covalent bonds.

Almost a similar effect was observed in the far-IR spectrum of the nanocomposite with 0.5 wt. % amino-MMT nanolayers. At higher amino-MMT loadings (1 wt.% and especially at 2 or 5 wt.%), i.e., for the nanocomposites with a higher degree of amino-MMT aggregation, some reverse changes in the far-IR spectrum were observed: the band of cycle vibrations displaced from 81 cm^{-1} to 78 cm^{-1} (Fig. 3.6b). This indicated the smaller influence of amino-MMT nanolayers on matrix dynamics, probably, due to their very inhomogeneous distribution in the matrix and formation of large agglomerates. However, a small suppressive influence on network skeletal torsional vibrations was observed after introducing 2 or 5 wt. % amino-MMT when the 194 cm^{-1} band displaced to 197 cm^{-1} (Fig. 3.6c).

Thus, the changes in the far-IR spectra do not contradict to the above TEM data providing the common idea about the structure-matrix dynamics interrelationship. The experimental data presented below demonstrate how the above peculiarities of nanostructure and matrix dynamics in the CER/amino-MMT nanocomposites are exhibited in their properties, including glass transition, relaxations, elastic properties, thermal stability, and creep resistance characteristics.

3.4 Dynamic Mechanical Analysis, Glass Transition

The peculiarities of the glass transition manifestation in the investigated nanocomposites were determined by the DMA and DSC methods. Figure 3.7 presents the dependencies of the mechanical loss factor, tanδ, on the temperature T, obtained for

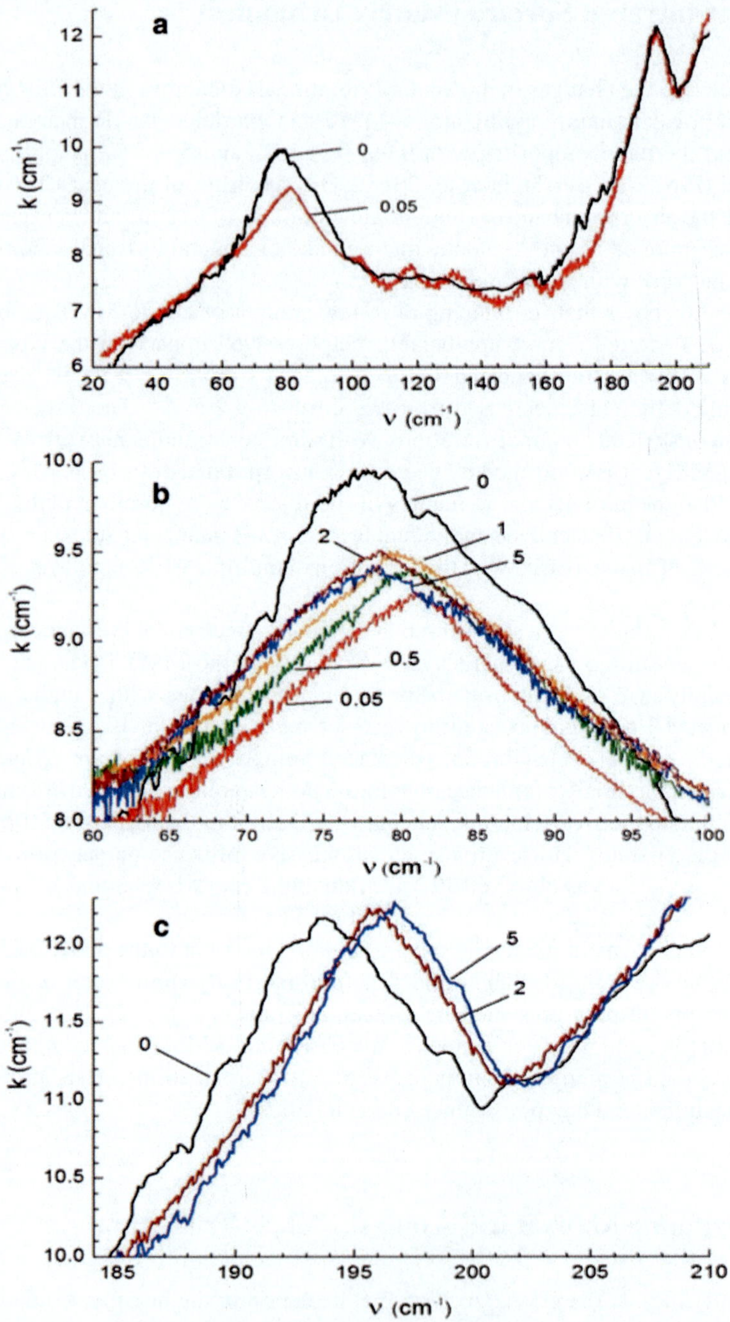

Fig. 3.6 Far-IR spectra of the neat DCBE-based CER matrix and the CER/amino-MMT nanocomposites. Numbers indicate amino-MMT contents in wt. %. **a** changing of the spectrum due to introducing 0.05 wt. % amino-MMT; **b** changing the absorption band of cycle vibrations in the nanocomposites; and **c** the absorption band of skeletal torsional vibrations in the spectra of neat matrix and two nanocomposites (Bershtein et al. 2015)

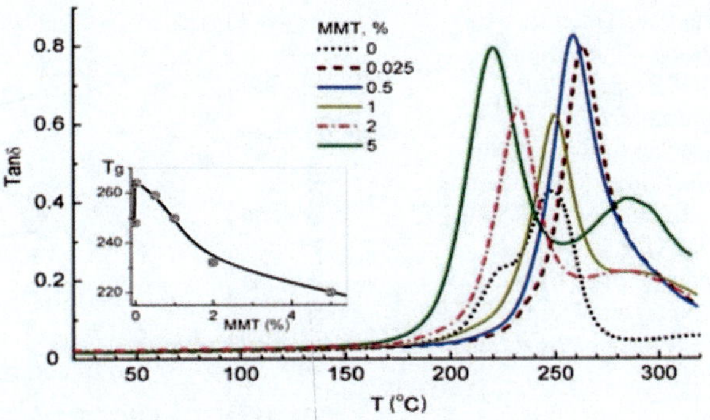

Fig. 3.7 DMA (1 Hz): temperature dependencies of tanδ for the DCBE-based CER matrix and the CER/amino-MMT nanocomposites with the indicated amino-MMT contents. Inset: T_g values taken as the temperatures of these peaks' maxima versus amino-MMT content plot (Bershtein et al. 2015)

a neat DCBE-based CER matrix and a series of CER/amino-MMT nanocomposites with 0.025, 0.5, 1, 2, and 5 wt.% nanofiller. The glass transition (α-relaxation) peak of the neat matrix with the maximum at T_g = 247 °C started from about 200 °C and completed at 270 °C. This peak is overlapping with the less intense peak (lower-temperature branch) at about 220 °C which may be attributed, obviously, to unfreezing dynamics in some of the matrix nanovolumes with non-completed cross-linking (network defects). It is a matter of principle that the largest positive effect from adding amino-MMT, including the substantial changes in the α–relaxation peak temperature position and disappearance of the above lower-temperature branch, was detected, again, already after embedding 0.025 wt.% amino-MMT units into the CER matrix. This led to an increase in the onset temperature of the glass transition peak by about 40 °C and achieved T_g = 263 °C. This, of course, was due to a good dispersity (exfoliating) and a more or less uniform distribution of the amino-MMT nanolayers in the matrix. Almost the same effect was achieved when 0.5 wt.% amino-MMT was introduced, but the opposite effect was observed when the amino-MMT content in the nanocomposites was increased to 1 wt.% and especially to 2 and 5 wt.%. In the latter case, T_g decreased down to 220 °C, and the glass transition peak started already from 180 °C (Fig. 3.7).

Additionally, Fig. 3.7 shows that in the relaxation spectra of nanocomposites with 1 wt.% or 2 wt.%, and especially 5 wt.% amino-MMT, the main peak is separated into two overlapping peaks. Along with the main glass transition peak, a less intense peak appears in the high-temperature region with a maximum at 280–290 °C. It should be assumed that the ambivalent effect of the nanofiller and the doublet nature (a peculiar bifurcation effect) of the glass transition in the higher nanofiller-loaded composites is due to the action of two factors. The presence of amino-MMT aggregates covalently grafted to the matrix in the composites obviously contributed both to the appearance of nanovolumes with cross-linking defects in the densely cross-linked polymer network (reduced T_g) and to nanovolumes undergoing strong constrained dynamics effects (anomalously high T_g values).

In other words, the latter is, obviously, a manifestation of the peculiar interfacial dynamics in the nanocomposites. At the same time, the relatively good exfoliation and uniform distribution of amino-MMT nanolayers in the matrix at their low content

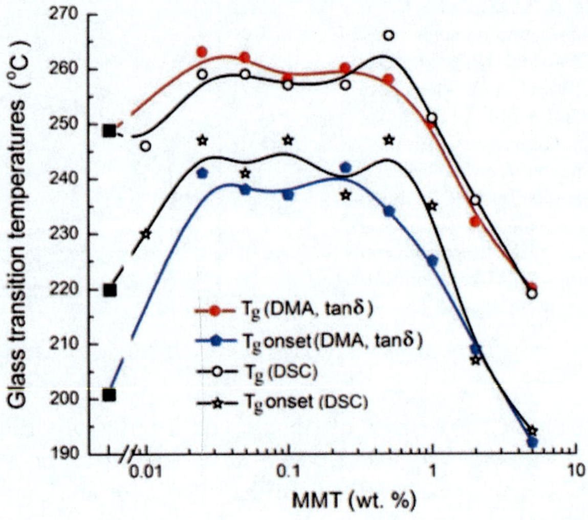

Fig. 3.8 Glass transition temperatures of the DCBE-based CER/amino-MMT nanocomposites as the function of MMT content (estimated by DSC and DMA). Black squares indicate glass transition temperatures for the neat CER matrix (Bershtein et al. 2015)

(<1 wt.%) contributed to the manifestation of the constrained dynamics effect in the matrix as a whole. As a result, this led to a shift of the single glass transition peak, including its beginning, to higher temperatures (Fig. 3.7).

In addition to evaluating the glass transition temperatures T_g's, at the maxima of the tanδ (T) curves and at the beginning of these relaxation peaks using DMA, these characteristic glass transition temperatures, T_g and T_g onset, were also measured by DSC for the DCBE-based CER/amino-MMT nanocomposites. Figure 3.8 presents these combined DMA/DSC data as the functions of amino-MMT content in the nanocomposites.

It can be seen that there is a general trend of changing the specified glass transition temperatures with a variation of the amino-MMT content in these nanocomposites. It is characterized by attaining the maximum values of the glass transition temperatures, T_g and T_g onset, already at ultra-low amino-MMT content (0.025 wt.%). The maximum increase, in the limit by 40 °C, occurred at the beginning of the glass transition. Practically the same positive effect was registered at different nanoparticle contents below 1 wt. %. However, the higher levels of amino-MMT loadings led to substantial negative effects, apparently due to the aggregation of amino-MMT nanoparticles. Thus, at 5 wt.% of nanoparticles, the glass transition temperature of the nanocomposite was by 30 °C lower than that of the neat matrix (Fig. 3.8).

Further, Fig. 3.9 shows the values of dynamic modulus E' (DMA, 1 Hz) obtained at 20° and 200 °C for the CER/amino-MMT nanocomposites as a function of nanofiller content.

A substantial increase in modulus values was observed almost in all cases, but the complicated character of these dependencies may be seen. At 20 °C, the approximately two-fold increasing modulus was attained at all amino-MMT contents, and, again, this effect was registered already at ultra-low nanofiller content of 0.025 wt.%. Maximal $E' = 4$ GPa was attained for the nanocomposite with 0.5 wt.% amino-MMT.

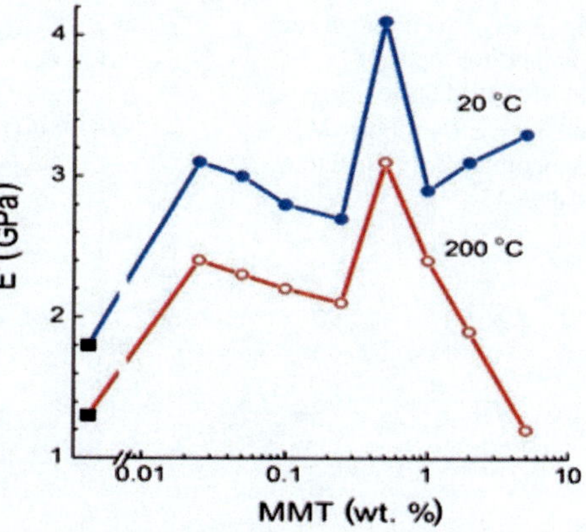

Fig. 3.9 DMA: dynamic (storage) modulus *E'* as a function of amino-MMT content in the DCBE- based CER/amino-MMT nanocomposites. Black squares indicate E' values for the neat CER matrix (Bershtein et al. 2015)

At 200 °C, i.e., in the vicinity of glass transition onset, nanocomposite modulus decreased, and the maximal positive effect was registered also at 0.5 wt.% amino-MMT when E' increased 2.7 times. Under these conditions, the superiority of the nanocomposite over the neat matrix disappeared completely at 5 wt. % amino-MMT (Fig. 3.9).

3.5 Creep Rate Spectra

Figure 3.10 presents the creep rate spectra obtained at two tensile stresses, 0.5 MPa and 0.05 MPa, for the neat matrix and the nanocomposites with 0.5, 1, 2, and 5 wt. % amino-MMT. The following points are noteworthy herein. First, the discrete character of these reproducible spectra, manifesting a few overlapping peaks within and in the vicinity of the glass transition range, is observed; their separation is more pronounced at the lower stress of 0.05 MPa (Fig. 3.10b). This is explained by the high resolution of the CRS technique (Bershtein and Yakushev 2010) and indicates the pronounced dynamic heterogeneity in the studied materials.

Secondly, again we can see the ambivalent influence of MMT on the obtained spectrum of creep rates, where the greatest positive effect of the growth of creep resistance is observed for nanocomposite with 1 wt.% amino-MMT. Indeed, if at 0.5 MPa appreciable creep of the neat matrix begins at 180 °C, a similar effect for this nanocomposite occurs only from 240 °C. As a result, the first sample fractures at 270 °C, whereas the second one is only at 370 °C (Fig. 3.10a). The opposite tendency, namely, the worsening of thermal and creep resistance properties of the CER matrix was observed in the creep rate spectra of nanocomposites containing 2 wt.% and especially 5 wt.% amino-MMT (Fig. 3.10a). In the latter case, the onset of a substantial creep shifted to 140 °C, and the rupture of the sample occurred at 250 °C.

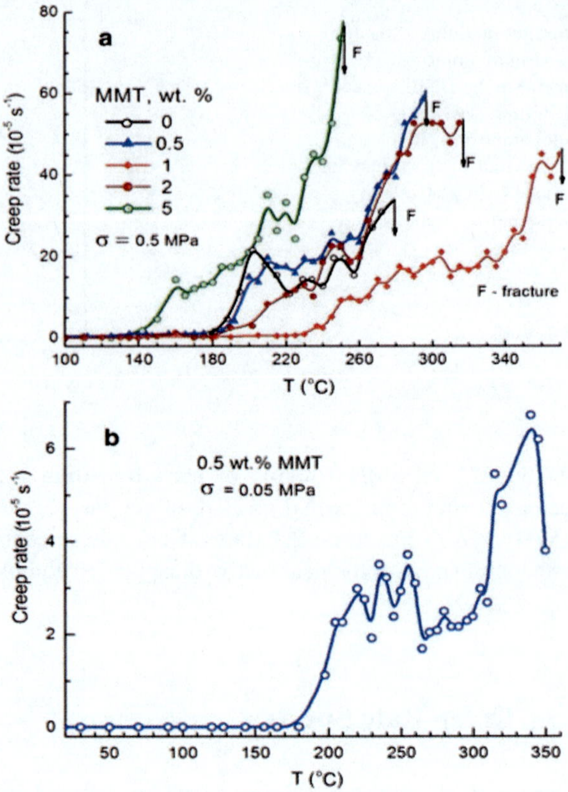

Fig. 3.10 Creep rate spectra obtained for the neat DCBE-based CER matrix and the CER/amino-MMT nanocomposites (Bershtein et al. 2015)

3.6 Thermal Stability

TGA investigation of non-isothermal decomposition of the CER/amino-MMT nanocomposites in comparison to the neat CER network has shown some influence of low amounts (<1 wt.%) of amino-MMT on the thermal stability of the nanocomposites (Fig. 3.11).

The degradation process of the neat CER started at ~ 420 °C corresponding to breakage of bonds between the phenyl and triazine rings of CER followed by intensive decomposition of cyanurate skeleton basically in the temperature range of ~ 440 to 700 °C. One can see that introducing low amounts of amino-MMT (0.01–0.5 wt.%) into the CER matrix resulted in a shift, by 10–15°, of the TGA curve to the higher temperatures. However, increasing amino-MMT content up to 5 wt.% in the nanocomposite resulted in a notable reduction of thermal stability that is in a good agreement with the above observations and explanations. Char residue was equal to 47–48 wt.% at 690 °C for all samples studied.

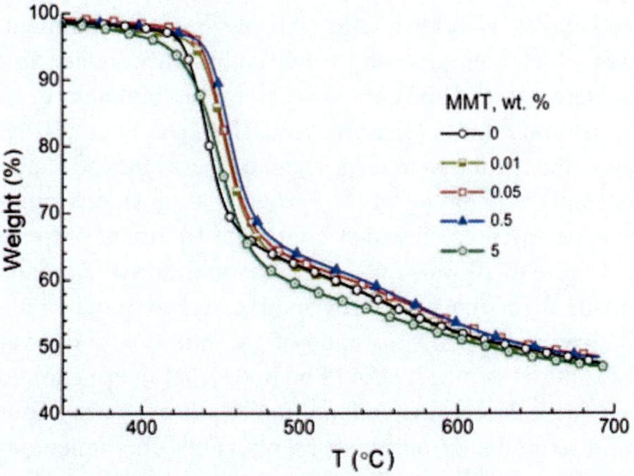

Fig. 3.11 TGA curves for the DCBE-based CER/amino-MMT nanocomposites (Bershtein et al. 2015)

3.7 Anomalous Composition and Properties of Micron Subsurface Layer in the CER-Based Nanocomposites

Baikova et al. (2016) obtained interesting experimental results in evaluating the composition and mechanical properties of the 1-micron subsurface layer of the CER-based nanocomposites. Three DCBA-based CER materials were investigated, namely a neat matrix and two nanocomposites containing 2.5 wt.% of either 3-D ECH-POSS nanoparticles or 2-D amino-MMT nanolayers. Their analysis was performed for the initial, untreated surface and after its treatment by stripping a ~ 50 μm thick surface layer by means of mechanical grinding and polishing.

Oxford instruments INCA energy dispersive X-ray (EDX) spectrometer was used for elemental analysis of subsurface layers. The layers of about 1, 1.5, and 3 μm thickness were analyzed using voltages of 3, 5, and 15 kV, respectively. A comparative estimation of the nanoparticle content in the above layers was performed by the determination of Si (for ECH-POSS) and Si and Al (for amino-MMT) content. The average values of element contents were determined from five EDX spectra (repeat measurements) in each case.

At the same time, the mechanical properties of the analyzed subsurface layers of the mentioned samples were evaluated by measuring microhardness at different loads, using a Vickers pyramid and PMT-3 instrument (LOMO, Russia). The microhardness was estimated at a loading time of 10 s and for indentation depths from 1 μm to 6 μm.

Table 3.1 presents the results of elemental (Si, Al) EDXS analysis of the subsurface layers of 1, 1.5, and 3 μm thickness for the two CER nanocomposites mentioned above. It can be seen that the content of Si and Al in the subsurface layers of nanocomposites is not constant and depends on the surface condition (initial or mechanically treated, with the removal of a 50 μm thick layer) and on the thickness of the analyzed layer. For the initial untreated surface, the Si and Al elements contents in the 1 μm thick layer turned out to be anomalously increased: twice as much as in the 3 μm thick layer. However, after the mechanical treatment of the nanocomposites, the anomalous composition of their subsurface layers disappears, and the nanoparticle element content does not depend on the thickness of the analyzed layer.

Microhardness measurements (Baikova et al. 2016) showed that the composition anomalies found in micron surface layers of these nanocomposites lead to anomalous mechanical behavior of the surface that must, obviously, be taken into account when considering mechanical strength and frictional properties of these nanocomposites.

Figure 3.12 presents the microhardness *H* versus indentation depth *h* plots obtained for the neat CER matrix and two analyzed nanocomposites. In general, one can see the inconstancy of the microhardness value at different values of the indentation depth. It should be noted that the phenomenon of increasing microhardness with decreasing indentation depth has been observed, especially in nano- and submicroindentation experiments. This phenomenon is known as the Indentation Size Effect (ISE) and has been widely studied, including for polymers and polymer nanocomposites (Diez-Pascual et al. 2015; Han 2011). This effect has been associated with indenter shape, frictional forces, dislocation models, etc.

Figure 3.12a shows relatively moderate ISE for the neat CER matrix when the microhardness *H* increases from 0.3 to 0.6–0.8 GPa with a variation of *h* from 5 μm to 1 μm; it was true both for the initial surface and after its mechanical treatment. This was not the case for nanocomposites.

On the one hand, almost the same curves (ISE) were obtained for both investigated nanocomposites with mechanically removed surface layer (Fig. 3.12b and c). It means that microhardness of internal layers of the nanocomposites was not changed in the presence of nanoparticles. However, ISE considerably increased, and the picture changed completely when measuring on nanocomposite samples with the initial untreated surface containing elevated contents of nanoparticles.

Figure 3.12b shows that when 3-D POSS nanoparticles are introduced into the matrix, the microhardness at the indentation depth of 1 μm increases to 1 GPa. The influence of introduced anisotropic nanoparticles—2-D MMT nanolayers (Fig. 3.12c)—was much greater: in this case, the microhardness *H* in the surface layer

Table 3.1 The results of elemental EDX analysis of subsurface layers in the DCBA-based CER nanocomposites with 2.5 wt.% ECH-POSS or amino-MMT nanoparticles (Baikova et al. 2016)

Sample	Surface state	Voltage (κV)	Layer thickness (μm)	Element content (at. %)	
				Si	Al
CER/ECH-POSS	Untreated	3	1.0	0.57	–
CER/ECH-POSS	Untreated	5	1.5	0.46	–
CER/ECH-POSS	Untreated	15	3.0	0.28	–
CER/ECH-POSS	Treated	5	1.5	0.24	–
CER/ECH-POSS	Treated	15	3.0	0.26	–
CER/amino-MMT	Untreated	3	1.0	0.55	0.21
CER/amino-MMT	Untreated	15	3.0	0.28	0.12

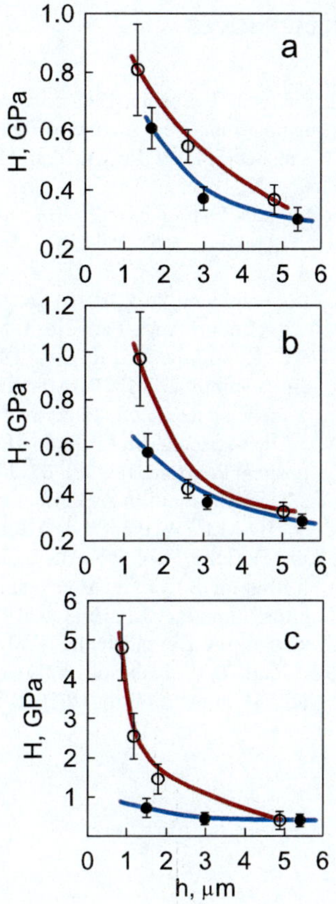

Fig. 3.12 Microhardness of the neat DCBA-based CER matrix **a** and the CER-based nanocomposites with 2.5 wt.% of ECH-POSS nanoparticles **b** or amino-MMT nanoparticles **c** as a function of indentation depth. The curves with white circles relate to the initial untreated surfaces, whereas black circles indicate microhardness values after mechanical grinding and polishing (Baikova et al. 2016)

of 1 μm thickness increased manifold: from 0.3 GPa (for inner layers) to 5 GPa. Simultaneously, additional embrittlement of the surface took place in this case since microcracks near indentations place were distinctly observed.

Thus, a combined microhardness study and EDX analysis of the surface layers of densely cross-linked CER network and two hybrid nanocomposites based thereon have been performed. An anomalously high ISE was found for a micron subsurface layer of the nanocomposites, which was most pronounced in the case of using a 2-D MMT nanofiller. It was shown that the elevated microhardness of this layer resulted from a large increase of nanoparticle content in the layer. Thus, it was shown that the particular composition of a micron-thick subsurface layer in the nanocomposites may be one of the reasons for the pronounced ISE in these materials. It should be recognized that the formation of the anomalous subsurface micron layer should be taken into account in some cases when considering the mechanical strength and frictional properties of these composites, especially for brittle composites (in which the fracture process starts from the surface).

References

L. Baikova, T. Pesina, D. Sakseyev, A. Fainleib, V. Bershtein, Hybrid cyanate ester resin-based nanocomposites: increased indentation size effect due to anomalous composition of micron subsurface layer. Polym. Test. **53**, 15–18 (2016). https://doi.org/10.1016/j.polymertesting.2016.05.001

F. Bergaya, G. Lagaly (eds), *Handbook of Clay Science* (Elsevier, 2013)

V.A. Bershtein, P.N. Yakushev, Laser-interferometric creep rate spectroscopy of polymers, in *Advances in Polymer Science*, vol. 230, ed. by K. Dusek (Springer, Heidelberg, Dordrecht, London, New York, 2010) pp. 73–219. https://doi.org/10.1007/978-3-642-13532-3

V.A. Bershtein, A.M. Fainleib, L.M. Egorova, K.G. Gusakova, O.P. Grigoyeva, D.A. Kirilenko, S.G. Konnikov, V.A. Ryzhov, P.N. Yakushev, N. Lavrenyuk, The impact of ultralow amounts of amino-modified MMT on dynamics and properties of densely crosslinked cyanate ester resins. Nanoscale Res. Lett. **10**, 165–179 (2015). https://doi.org/10.1186/s11671-015-0868-5

A.M. Diez-Pascual, M.A. Gomez-Fatou, F. Anis, A. Flores, Nanoindentation in polymer nanocomposites. Progr. Mater. Sci. **67**, 1–94 (2015)

C.-S. Han, Indentation size effect in polymers, in *Advanced Materials Science Research*, vol. 10, ed. By M.C. Wythers (Nova Science, New York, USA, 2011), pp. 393–413 (Chapter 15)

Y. Lin, M. Song, C.A. Stone, S.J. Shaw, A comprehensive study on the curing kinetics and network formation of cyanate ester resin/clay nanocomposites. Thermochim. Acta. **552**, 77–86 (2013). https://doi.org/10.1016/j.tca.2012.11.009

L. Nicolais, A. Borzacchiello, S.M. Lee (eds.), *Encyclopedia of Composites* (Wiley, 2012)

D.R. Paul, L.M. Robeson, Polymer nanotechnology: nanocomposites. Polymer **49**, 3187–3204 (2008). https://doi.org/10.1016/j.polymer.2008.04.017

Chapter 4
Nano- and Subnanocomposites with Silica Units Introduced by a Sol–gel Method

Abstract The results of physical studies of the hybrid CER-silica composites with 0.01–10 wt.% silica prepared via a sol–gel technology are presented. HAADF-STEM, EDXS, DMA, DSC, and TGA were used for their characterization. It was found that the greatest positive effect of introducing covalently embedded silica units into the matrix is observed at their ultra-low content of 0.02–0.1 wt.%. The possibility of creating subnanometer-sized silica nodes in the matrix network in the absence of clustering was revealed in this case, and the term "subnanocomposites" was introduced into the literature. The effect of the quasi-regular distribution of silica subnanounits or molecularly dispersed nanoparticles in nanovolumes of amorphous matrix was detected. The superiority of the properties of polymer subnanocomposites over those of the corresponding composites with SiO_2 nanoclusters was shown. Certain correlations between the nanostructure, matrix dynamics, and the properties of the investigated composites were traced. Additional calculations based on the obtained structural data were performed, which made it possible to establish the nature of the above effects. It turned out that the dynamics of the matrix as a whole transformed into the state of constrained interfacial dynamics in this case.

4.1 Structure

Jothibasu et al. (2011) and Devaraju et al. (2013) have performed the first syntheses and analyses of CER-silica hybrid nanocomposites using the sol–gel method. 4,4-dicyanato-2,2-diphenyl propane (Jothibasu et al. 2011) or 1,4-bis(2-(4-cyanatophenyl)-2-propyl)benzene monomers (Devaraju et al. 2013) were used to create the organic matrix in these nanocomposites. To create the silica phase in these nanocomposites, tetraethoxysilane (TEOS) was used, which was hydrolyzed followed by the condensation reaction of the silanol groups and the formation of a SiO_2 network. In addition, γ-aminopropyltriethoxysilane (γ-APTES) or 3-glycidoxypropyltrimethoxysilane (GPTMS) were used as coupling agents chemically embedded in the silica network, which ensured its covalent bonding with the organic matrix due to reaction between the cyanate ester (OCN) groups and amine

V. A. Bershtein and P. N. Yakushev, *High-Temperature Polymer Nanocomposites Based on Heterocyclic Networks from Nitrile Monomers*, Springer Series in Materials Science 334, https://doi.org/10.1007/978-3-031-32943-2_4

or glycidyl groups. It should be emphasized that the nanocomposites with a large content of inorganic fractions were investigated in the works (Devaraju et al. 2013; Jothibasu et al. 2011). Although the SiO_2 content was not specified by the authors, the cyanate ester monomer and TEOS were taken in a 1:1 weight ratio, along with varying molar equivalents (0.5, 1.0, 1.5, and 2.0) of coupling agents γ-APTES or GPTMS with respect to the monomer (Devaraju et al. 2013).

FTIR spectra (Devaraju et al. 2013) confirmed the expected molecular structure of the investigated hybrid nanocomposites. The completeness of matrix polymerization was confirmed by the disappearance of the peak at 2237 cm^{-1}, corresponding to the –OCN group of cyanate ester. The appearance of the peaks at 1365 cm^{-1} and 1565 cm^{-1} indicated the appearance of triazine cycles in the matrix structure, and the appearance of the peak at 1100 cm^{-1} confirmed the formation of a silica network. The occurrence of a peak at 1604 cm^{-1} indicated the hybridization reaction between the –OCN group and the amino group of γ-APTES. Finally, the reaction between the –OCN groups of cyanate ester and the glycidyl group of GPTMS was confirmed by the disappearance of the band at 914 cm^{-1} and the appearance of a band at 1678 cm^{-1} corresponding to an oxazoline linkage. Figures 4.1 and 4.2 present schematically the molecular structures of CER-SiO_2 hybrid nanocomposites using γ-APTES or GPTMS as coupling agents (Devaraju et al. 2013). According to SEM estimates, the SiO_2 particles in the hybrid films in question were about 5 nm in size.

In (Devaraju et al. 2013; Jothibasu et al. 2011) it was found for the first time that the incorporation of silica units into the cyanate ester matrix led to an improvement in its thermal properties. Thus, the DSC data showed that the glass transition temperature T_g values for CER-silica nanocomposites were significantly higher than for neat matrix as a consequence of the formation of covalent bonds between the matrix and nanoparticles, which led to significant suppression of the matrix molecular mobility. The best result in thermal degradation resistance was achieved using APTES: in this case, the mass loss of CER-SiO_2 nanocomposite at 400 °C decreased from 80 to 20%, and the char yield at 900 °C increased from 7 to 46% relative to the characteristics of the neat matrix (Devaraju et al. 2013). It was also shown in this work that the dielectric constant decreased from 3.33 for the neat matrix to 2.40 for the nanocomposite, which is also significant in some of its applications.

In recent years, Bershtein et al. (2016a, b, c, 2021a, b) have published the results of an extensive physical study of CER-silica nanocomposites. In these works, hybrid CER-silica nanocomposite films were formed in situ using a sol–gel technique from 1,1-bis(4-cyanophenyl)ethane (dicyanate ester of bisphenol E (DCBE), TEOS, and γ-aminopropyltrimethoxysilane (APTMS) as a coupling agent. Different compositions were synthesized with a ratio of TEOS:APTMS = 90/10 wt.% and with the estimated silica contents of 0.01, 0.025, 0.05, 0.1, 0.25, 0.5, 1, 2, 5, and 10 wt.%.

In the publications (Bershtein et al. 2016a, b, c, 2021a), the main reactions of the hybrid composite formation were also confirmed by FTIR, including the reaction of polycyclotrimerization of cyanate ester with the formation of triazine cycles; hydrolysis of TEOS and APTMS with the subsequent polycondensation of the resulting silanol products, as well as chemical hybridization of organic and inorganic components due to the interaction of -OCN groups with NH_2-groups of APTMS. Figure 4.3

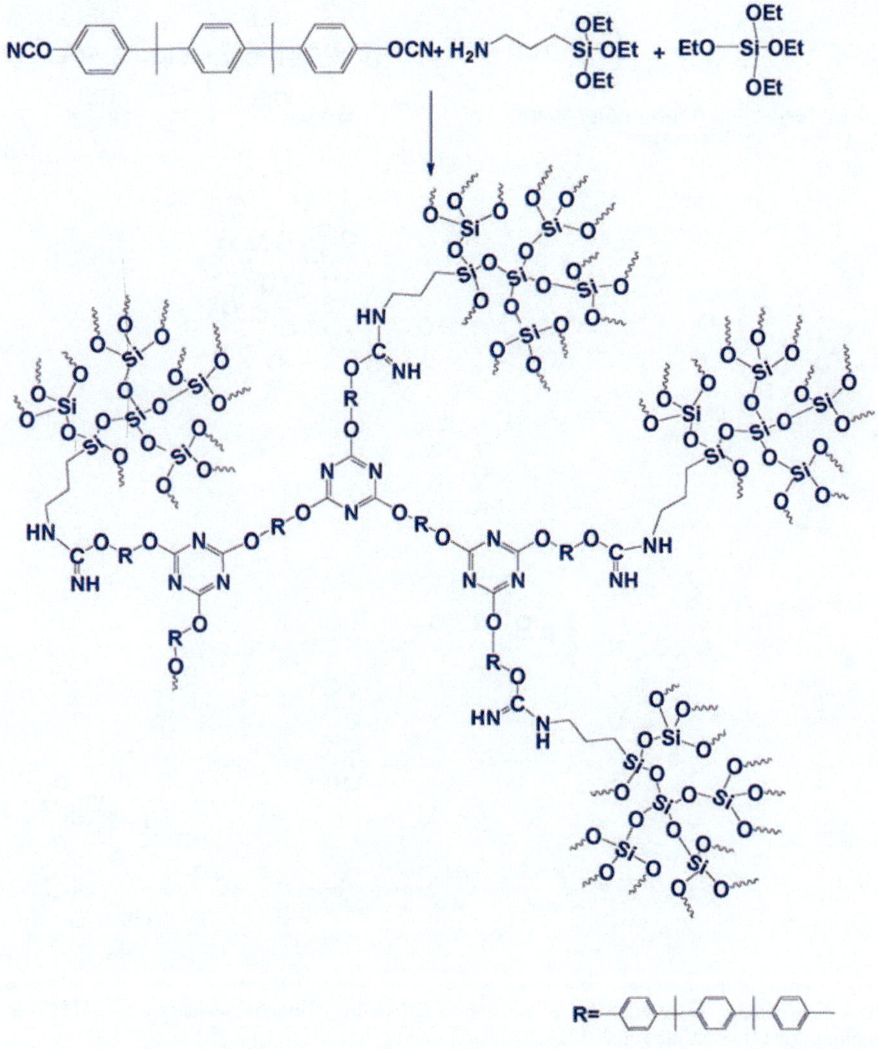

Fig. 4.1 Schematic representation of CER-silica hybrid nanocomposite using APTES as coupling agent (Devaraju et al. 2013)

shows the scheme of the CER-silica composite network with a covalently embedded silica node.

Figure 4.4 shows the results of the structural analysis of several composites studied: HAADF-STEM images (**a–e**) and Si nanodistribution maps (on the right). At 0.1% SiO_2 no silica clusters are observed in the composite. This means that within the noise level of HAADF-STEM images, the composite is homogeneous. The absence of nanoheterogeneity suggests that the matrix network of this composite may contain

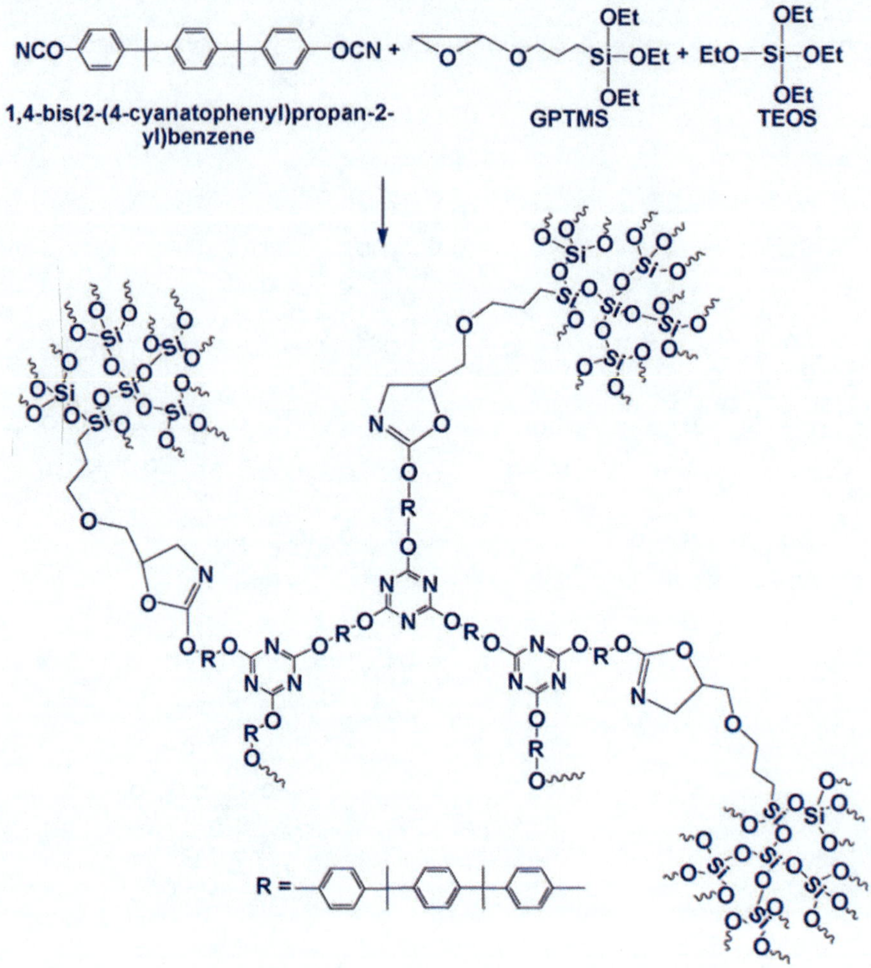

Fig. 4.2 Schematic representation of the hybrid CER-silica nanocomposite using GPTMS as coupling agent (Devaraju et al. 2013)

only silica "nodes" of subnanometer size. The scheme of such a node is shown in Fig. 4.4a; this point is discussed below.

The beginning of the nanoclusterization in the composite is observed only with the introduction of 0.5% SiO_2 when clusters 2–4 nm in size become distinguishable. In the composite with 1% SiO_2, individual, sparsely arranged nanoclusters of 20–40 nm in size are observed, and already at 2% SiO_2 in the composite aggregates of nanoclusters are formed. Their size increases with increasing silica content. Thus, in composites with 2% and 10% SiO_2 aggregates of about 100 nm and (0.5–1.0) μ are found, respectively (Fig. 4.4d, e). As the EDXS maps of Si nanodistribution in

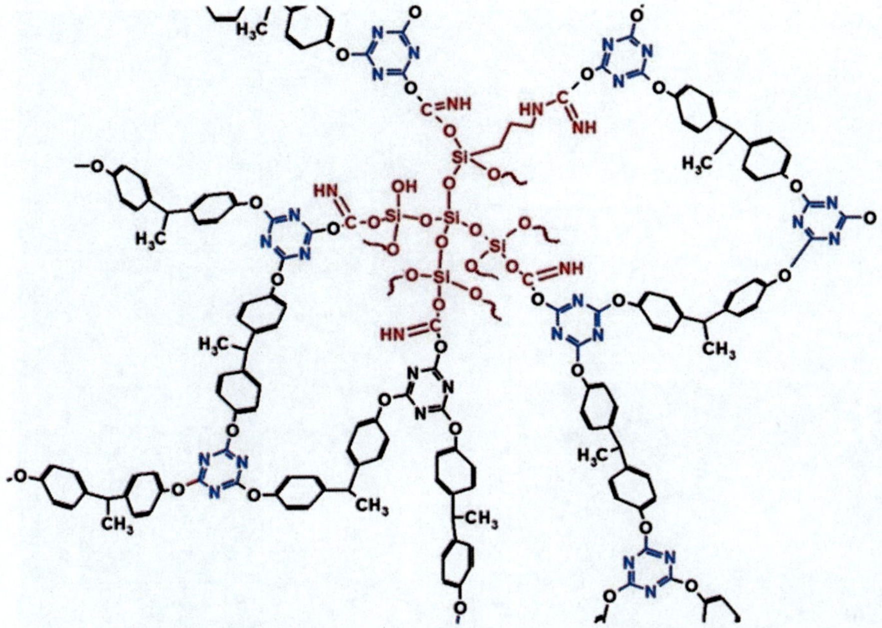

Fig. 4.3 Scheme of CER-silica hybrid network with covalently embedded silica node (Bershtein et al. 2016a)

the composites with 1, 2, and 10% SiO_2 presented here show, the observed structural heterogeneity is definitely due to the presence of silica clusters and their aggregates.

Figure 4.5 shows typical EDX spectra obtained for 1 × 1x100 nm^3 volumes and the differences in measured Si characteristic peaks in three cases: when the electron beam was directed to a single silica cluster, to an aggregate of clusters, and to a matrix, or to the composite with 0.1% SiO_2. A large difference in the intensities of the Si peaks in these cases is evident. The small Si peak in the spectrum of the matrix is explained by some "hitting" of SiO_2 units in the electron beam path.

The most interesting question is the formation of silica units when their content in the composite is very low, in particular at 0.1% SiO_2. On the one hand, such a composite is nanohomogeneous and contains no clusters. However, on the other hand, Si-containing compounds were introduced and EDX spectra taken many times from different points of this sample invariably confirmed in all cases the presence of Si in all of the above nanovolumes (it can be illustrated for example by the spectrum given in the log-scale plot of the inset in Fig. 4.5).

To clarify this question, it should be taken into account that TEOS and APTMS are very prone to hydrolysis reactions with subsequent polycondensation of the resulting silanol products at moderate temperatures to form silica structures. In addition, as shown below, it was in this case that the silica units had the maximum positive effect on the properties of the CER matrix. The above gave natural grounds to assume that silica units can form in this case only additional inorganic nodes of subnanometer size

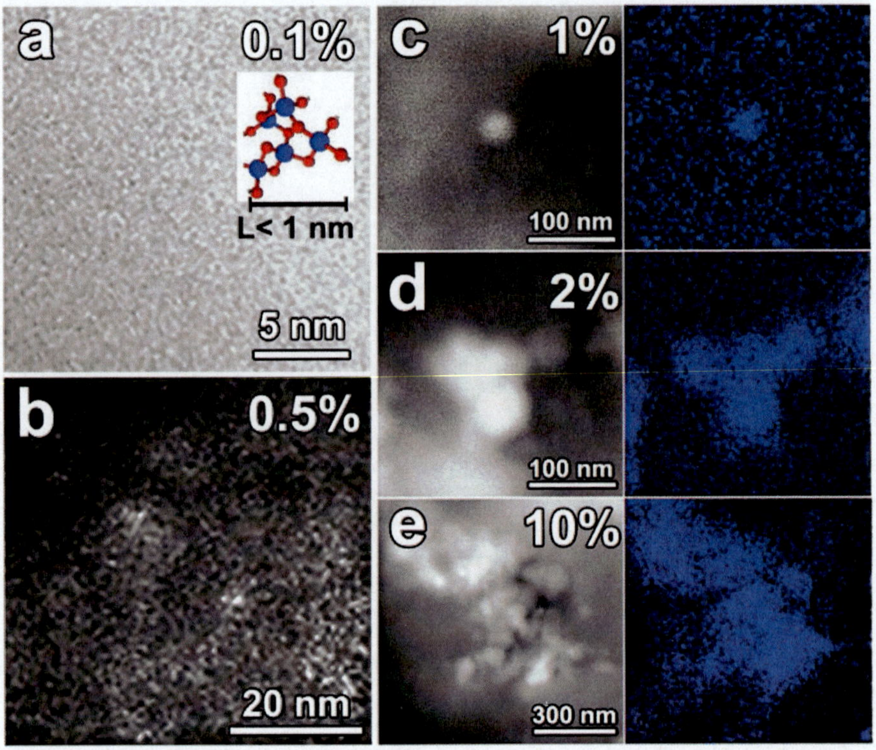

Fig. 4.4 HAADF-STEM images (**a–e**) obtained for the DCBE-based CER-silica composites with SiO_2 contents of 0.1, 0.5, 1, 2, and 10 wt. %, and Si nanodistribution maps (on the right) obtained by EDXS for the composites with 1, 2, and 10 wt. % SiO_2 (Bershtein et al. 2016a)

in the polymer matrix network, and these nodes are distributed more or less uniformly over the matrix volume. The size of such a node L should be in the interval: 0.5 nm $\leq L < 1$ nm. The lower limit for the value of L is determined by the Si–O bond length of 0.16 nm, while the upper limit is obvious: the HAADF-STEM technique recognizes silica units of 1 nm and larger. A subnanometer-sized silica node is shown schematically in Fig. 4.4a.

It should be noted that in the hypothetical case of a completely uniform, regular distribution of silica nodes in the matrix, an "elementary" cubic volume with a 5–10 nm side would include one silica node, and the average distance between the nodes would also be 5–10 nm.

Such a structure (the composite with ultra-low silica content) could be considered as a polymer network with embedded foreign inorganic units less than 1 nm in size. However, the authors preferred to consider it and name it a polymer subnanocomposite. This term seems to have been first introduced into the polymer literature with respect to such systems; to date, it has been successfully used by us in five publications (Bershtein et al. 2016a, b, c, 2021a, b). Moreover, there are experimental

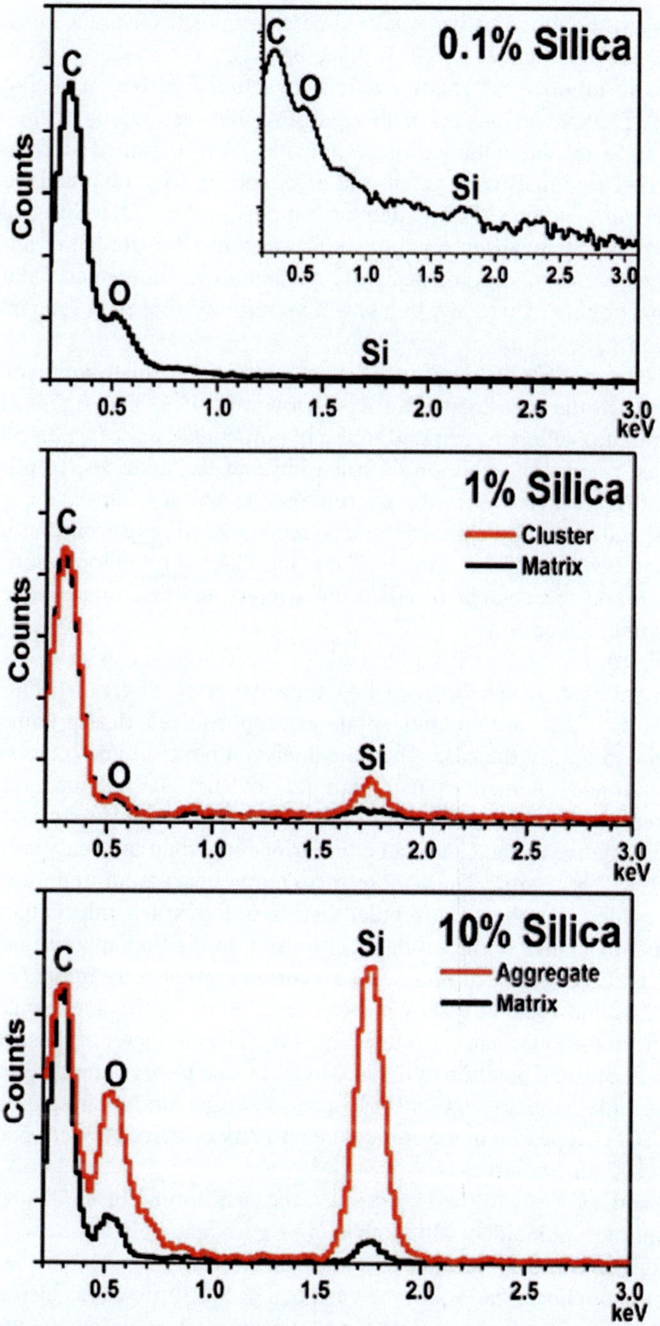

Fig. 4.5 EDX spectra of the DCBE-based CER-silica nanocomposites with 0.1% SiO_2 (no clusters, Si presence may be discerned in the log-scale plot, see inset), 1% SiO_2 (inside silica cluster and in the matrix), and 10% SiO_2 (inside silica cluster aggregate and in the matrix). Nanovolumes of 1 × 1 × 100 nm^3 in size were analyzed (Bershtein et al. 2016a)

grounds to assert the superiority of such subnanocomposites over the corresponding nanocomposites (Bershtein et al. 2021a) (see below).

In order to accumulate information about the actual distribution of silica units in CER matrices EDXS was used. For this purpose, the histograms of the statistical distribution of Si in nanovolumes of composites were obtained (Bershtein et al. 2016a, 2021a). The signals were taken, and EDX spectra were obtained mainly from $1 \times 1 \times 100$ nm^3 volumes. For a subnanocomposite with 0.1% SiO_2, histograms were also obtained by means of obtaining spectra from $10 \times 10x5$ nm^3 and $2 \times 2 \times 5$ nm^3 volumes corresponding respectively to the above "elementary" nanovolume and part of this volume. Up to 300 points were randomly chosen in each material for the EDX spectra acquisition.

Figure 4.6 shows eight histograms of the actual nanoscale distribution of Si atoms, i.e., silica units in the investigated CER-silica composites. This figure also shows Gaussian contours, which correspond to the hypothetical case of a perfectly homogeneous, quasi-regular distribution of silica units in the matrices. In other words, these are contours whose width (the instrumental function width) is determined by the stochastic nature of the characteristic X-ray signal arising from the interaction of the electron beam with the sample. Thus, the Gaussian contours correspond to the situation when the content of silica units in all analyzed nanovolumes of the composite is unchanged.

In this regard, the histograms obtained for the composite with 0.1% SiO_2 are of particular interest. It can be seen that the histogram of Si distribution in the volumes $1 \times 1 \times 100$ nm^3 of this subnanocomposite practically coincides with the Gaussian contour. In the case when the analyzed nanovolumes corresponded to the abovementioned "elementary nanovolumes" of $10 \times 10 \times 5$ nm^3, only a weak deviation from the Gaussian contour is noted. However, a significant broadening of the histogram relative to the Gaussian contour occurs when the analyzed volume is reduced to $2 \times 2 \times 5$ nm^3. Thus, these histograms provide an important physical result: clear evidence of the quasi-regular distribution of silica units in nanovolumes of the amorphous matrix in the subnanocomposite, in the first approximation.

At the same time, the occurrence of structural heterogeneity in nanocomposites with 0.5, 1, 2, 5, and 10% SiO_2 shown above resulted in significant deviations of the histograms from the Gaussian contours (Fig. 4.6). The character of these deviations, including the shape and position of the maxima of these histograms, differed dramatically when single nanoclusters (at 0.5% and 1% SiO_2), and their aggregates (at 2, 5, and 10% SiO_2) appeared in the composites (the measurements were performed in the $1 \times 1 \times 100$ nm^3 volumes).

At 0.5% and 1% $Si0_2$ in the composites, the maximums of the histograms and Gaussian contours coincided, although their significant divergence was observed in the part of increased Si content in some nanovolumes as a consequence of the beginning of nanoclusterization. In the case of 1% SiO_2, there was also a tendency for the appearance of "silicon-depleted" nanovolumes. A complete mismatch of the histograms with the Gaussian contours was observed in the cases of the formation of cluster aggregates in the composites. It can be seen that at 2, 5, and 10% SiO_2 in the composites, there is a sharp broadening of the histograms, they become asymmetrical,

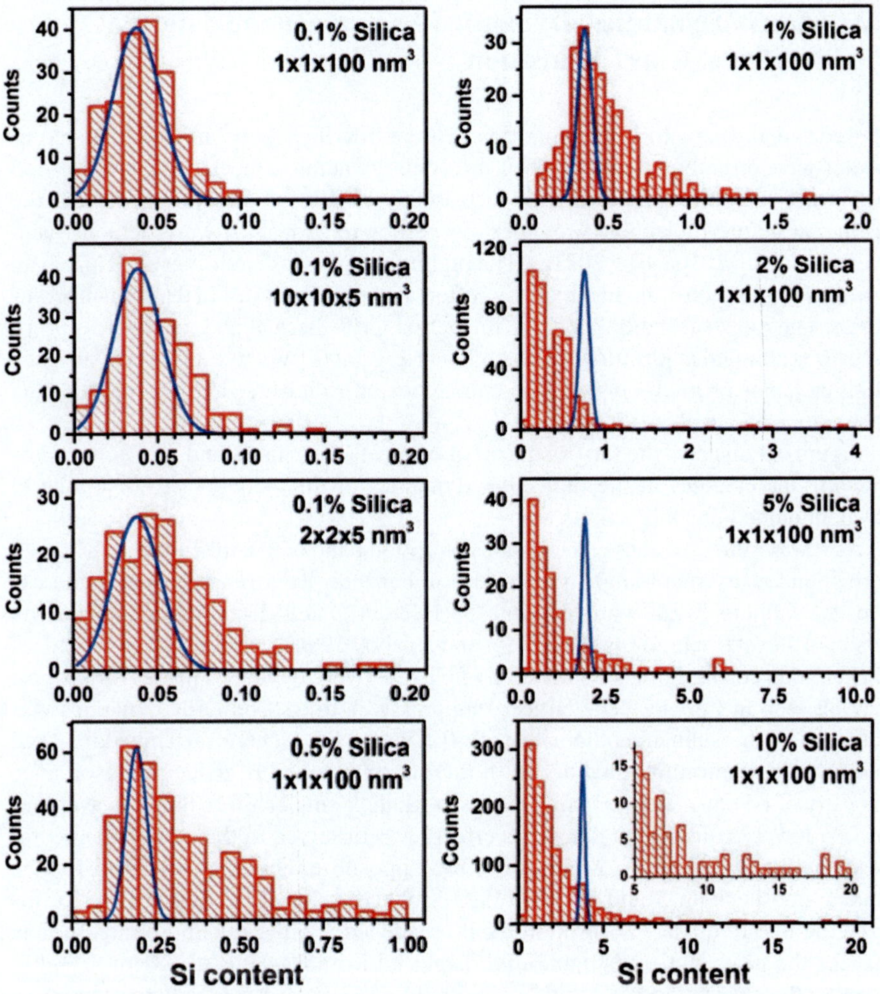

Fig. 4.6 Histograms of Si content (at. %) in the indicated nanovolumes for the DCBE-based CER-silica nanocomposites containing 0.1, 0.5, 1, 2, 5, and 10 wt. % SiO_2, as estimated by EDXS. Blue Gaussian contours correspond to regular silicon nanodistribution in the composites (Bershtein et al. 2016a)

and anomalously located maximums of histograms are observed in sharply silica-depleted regions. Thus, for example, at a mean SiO_2 content of 10 wt.% (Si content of about 3 at.%) the real Si content in nanovolumes varied from a value much smaller than 1 at.% (with a maximum in this region) to 20 at.%. It means that the sharply silica-depleted nanoregions prevail and coexist with the highly silica-enriched nanoregions in this nanocomposite.

4.2 Matrix Dynamics, Dynamic Mechanical and Thermal Analyses, Glass Transition

The considered structural peculiarities of the CER-SiO_2 subnano- and nanocomposites were directly reflected in their molecular dynamics, mechanical and thermal properties, as controlled by far-IR spectroscopy, DMA and DSC methods, and the distinct structure-molecular dynamics–properties relationships were demonstrated (Bershtein et al. 2016a, b, c, 2021a). The highlight of this research consisted in that an extremely high "constrained dynamics" effect was registered just in the subnanocomposite. The most substantial positive impact on CER thermal and mechanical properties was attained at ultra-low silica contents, e.g., at 0.1 wt. % SiO_2, in the absence of silica nano- or microclusters; the subnanocomposite exceeded in properties the nanocomposites with 1–10 wt.% SiO_2.

Figure 4.7 displays the far-IR spectra of the neat CER matrix and four composites showing the changes in the molecular dynamics of this network at 20 °C due to covalent embedding silica units.

As it was indicated above, the broad absorption band with a maximum at 81 cm^{-1} corresponded to small-angle vibrations of benzene and triazine rings, whereas two overlapping bands with maxima at 193 cm^{-1} and 230 cm^{-1} characterized torsional skeletal vibrations (Bershtein and Ryzhov 1994). Two opposite tendencies of changing, constraining, or activation of CER network dynamics in the composites, may be seen in Fig. 4.7. The largest "suppressive" impact on CER dynamics was observed in the subnanocomposite with 0.1 wt.% SiO_2: chemical embedding and uniform distribution of silica nodes in the matrix resulted in reducing absorbance intensity of 81 cm^{-1} and 193 cm^{-1} bands, with their small shift to higher wavenumbers. A lower "constrained dynamics effect" was observed in the case of nanocomposite with 1 wt.% SiO_2. Meanwhile, the opposite effect was registered for the nanocomposite with 5 and 10 wt.% SiO_2-containing clusters' aggregates. In this case, the low-frequency absorption band broadened and became more intense than that for the neat matrix; its maximum displaced from 81 cm^{-1} to 78 cm^{-1}. Additionally, the absorption band with a maximum of 230 cm^{-1} was transformed into a doublet of 230 and 215 cm^{-1} bands. The latter changes indicated some intensification ("facilitating") of network dynamics because of its elevated defectiveness and enhanced structural and dynamic heterogeneities.

Such ambivalent influence of introduced silica units totally manifested itself in the thermal, relaxation, and elastic properties of composites (Figs. 4.8 and 4.9). One can see that T_g as the maximum in the tanδ (T) plot and modulus E' increased by 50° and from about 2 GPa to 3 GPa, respectively, in the subnanocomposite with 0.1 wt.% SiO_2. Contrarily, the formation of silica nanoclusters and especially their aggregates of hundreds of nanometers in size at silica contents of 2–10 wt.% led to a distinctly negative impact on the matrix properties. At 10 wt.% SiO_2, the pronounced structural heterogeneity resulted in the large decreasing thermal properties and revealed in three glass transition temperatures (DSC data) in the nanocomposite (Fig. 4.9).

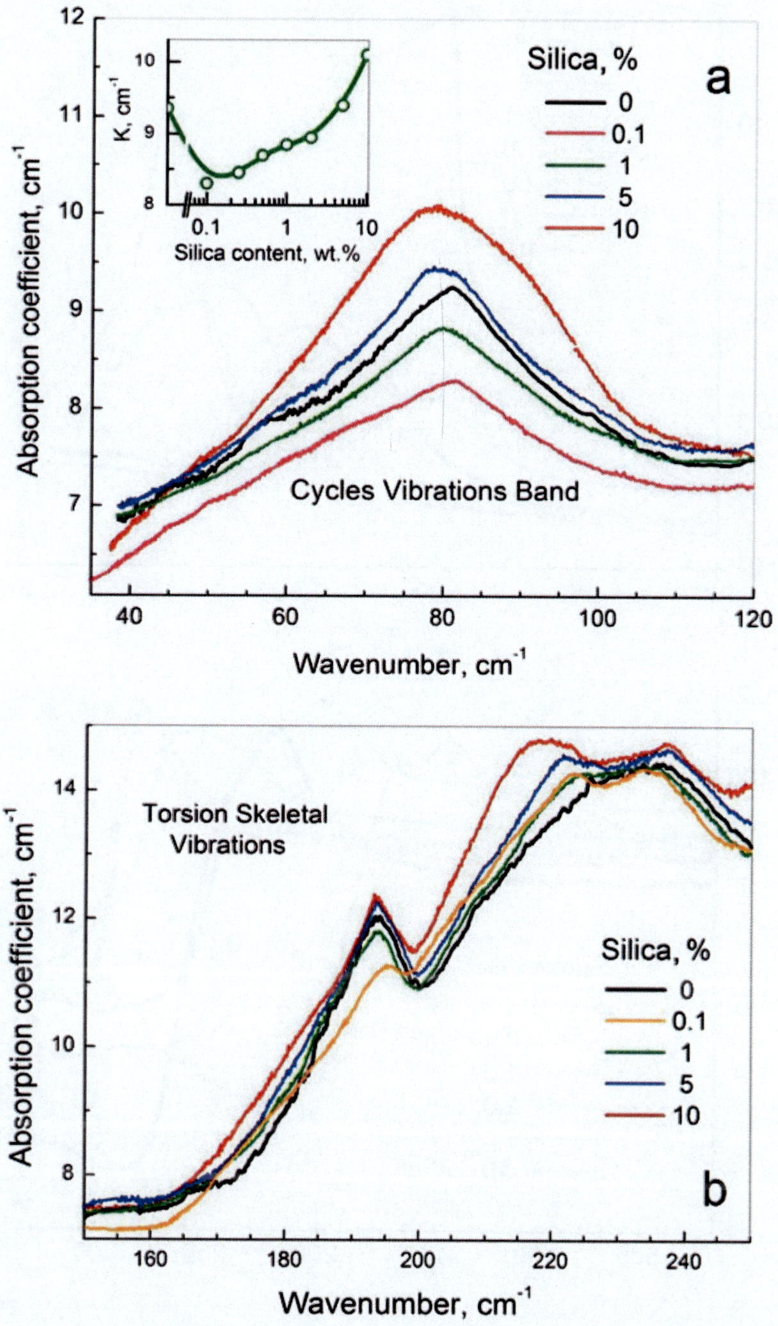

Fig. 4.7 Far-IR spectra of the neat DCBE-based CER matrix and CER-silica nanocomposites with 0.1, 1, 5, and 10 wt.% SiO_2 in the spectral regions of the absorption bands of **a** cycles' vibrations and **b** torsional skeletal vibrations in the CER network. Inset: the dependence of the absorption coefficient of a low-frequency band on silica content in the nanocomposites (Bershtein et al. 2016a)

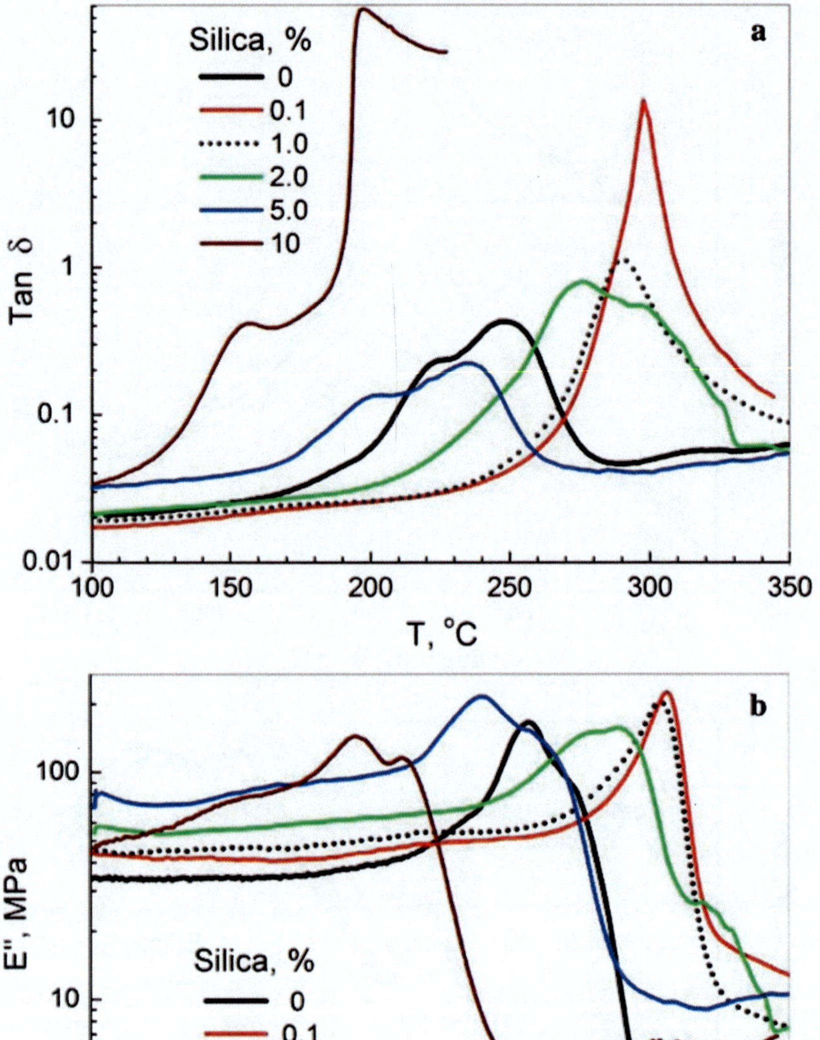

Fig. 4.8 DMA (1 Hz): **a** mechanical loss factor tanδ (T) and **b** dynamic (storage) modulus E'(T) plots obtained for the neat DCBE-based CER matrix and CER-SiO_2 nanocomposites with 0.1, 1, 2, 5 and 10 wt. % SiO_2 (Bershtein et al. 2016a)

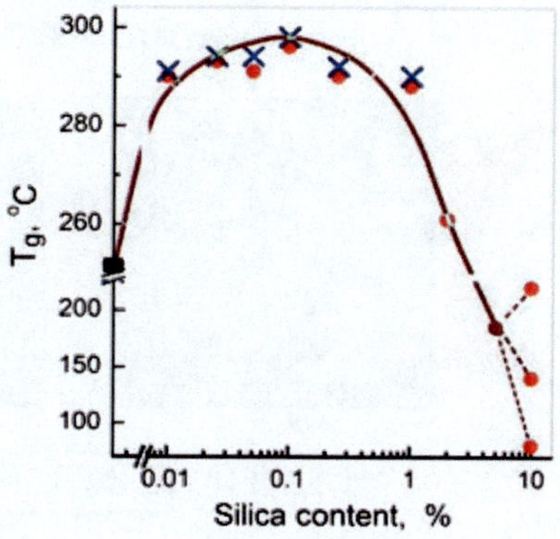

Fig. 4.9 Glass transition temperatures of the CER-silica composites versus composition plots as estimated by DSC (circles) and DMA (crosslets). Black square designates *T*g for the neat matrix (Bershtein et al. 2016a)

Devaradju et al. (2013) performed TGA of the CER-SiO_2 nanocomposites where matrix network was formed from 1,4-bis(2-(4-cyanatophenyl)-2-propyl)benzene monomer (see Fig. 4.10). This cyanate ester did not provide the optimal thermal properties of the neat matrix: its T_g was equal to 182 °C only increasing up to 250 °C in the CER-SiO_2 nanocomposite (Fig. 4.9 shows that in our study (Bershtein et al. 2016a) *T*g was about 250 °C for the neat matrix and increased up to about 300 °C for the nanocomposite; its thermal degradation started in fact from 400 °C only).

Nevertheless, Fig. 4.10 demonstrates a large advantage in the thermal stability of the nanocomposites regarding the neat polymer network: for instance, heating the neat matrix to 400 °C resulted in mass loss of about 80% and char residue of 10% at 900 °C, whereas for the best nanocomposite mass loss at 400 °C was only 20% and char residue of 50% at 900 °C was obtained. $BACE_1$, $BACE_2$, $BACE_3$, and $BACE_4$ designations in Fig. 4.10 correspond to cyanate ester:APTES:TEOS ratios of 1:0.5:1, 1:1:1, 1:1.5:1, and 1:2:1 at the synthesis of nanocomposites.

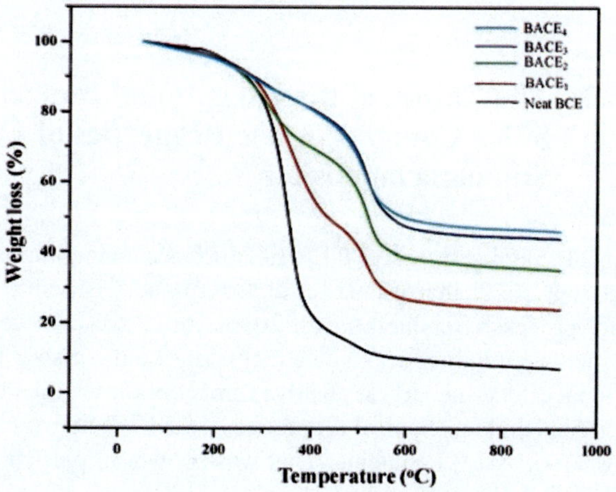

Fig. 4.10 TGA thermograms of neat CER matrix (BCE) and CER-SiO_2 (BACE, see text) hybrid nanocomposites using APTES as the coupling agent (Devaraju et al. 2013)

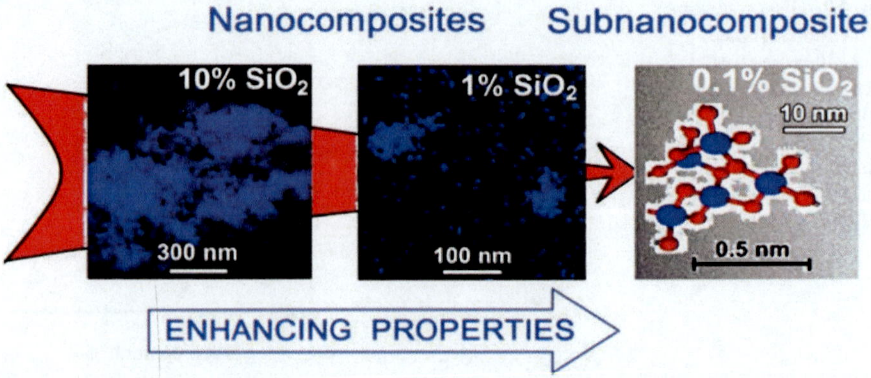

Fig. 4.11 Schematic illustration of enhancing properties of CER-SiO_2 composites in a row: samples with "large" SiO_2 nanocluster aggregates—samples with SiO_2 nanoclusters—samples with subnanometer-sized silica nodes in a polymer matrix. The silica node is shown here. Clusters were estimated using STEM images and EDXS technique (maps of Si nanodistribution)

In conclusion of this section, we give the illustrative Fig. 4.11 representing the nanostructures of the CER-SiO_2 composites with large, moderate, and ultra-low silica contents.

If the former two composites may be considered as nanocomposites with silica clusters and their aggregates, the latter one may be named, really, as a ***subnanocomposite***. Just only silica subnanometer-sized nodes, chemically embedded into a polymer amorphous network and distributed quasi-regularly in its nanovolumes, may presumably exist in this case. As a result, such structure and, probably, the enhanced "constrained dynamics effect", caused by SiO_2 nodes embedded into this densely cross-linked polymer network, provided the observed superiority of the properties of subnanocomposites.

4.3 The Nature of the Exceptional Impact of Ultra-Low Silica Contents on the Properties of CER-Based Subnanocomposites

In the publications on CER/ECH-POSS nanocomposites (Bershtein et al. 2015, 2016d, 2020; Starostenko et al. 2012) and CER-SiO_2 hybrids obtained via a sol–gel process (Bershtein et al. 2016a, b, c, 2021a), we have revealed not only the ambivalent influence of silica inclusions on the matrix properties but also the exceptional effect: the largest positive impact of these silica units on molecular dynamics, thermal and mechanical properties of the CER matrix was revealed at their ultra-low (0.03–0.1 wt.%) contents. That was shown, in particular, above in Figs. 2.16, 2.22, 2.25, 3.7, 3.8, 4.8, and 4.9.

Our research (Bershtein et al. 2021b) clarified the nature of the exceptional positive impact of the ultra-low content of silica (SiO_2) units on the properties of the heterocyclic cyanate ester resin (CER) network matrix in the subnano- or nanocomposites. To this effect, we have performed a series of calculations of average distances d between SiO_2 units and the interfacial area values S in the composites as functions of the content and size of SiO_2 units. The results of these calculations were analyzed in combination with the experimental structural data (Bershtein et al. 2016a, 2015) obtained for these composites.

Before this research, we supposed, a priori, that the exceptional positive influence of ultra-low silica contents might be associated with the substantial contribution of the special matrix dynamics at the interfaces in the studied composites. We took into account that several articles have recently discussed the problem of interfacial matrix dynamics in composites (Baeza et al. 2016; Cheng et al. 2017; Berriot et al. 2002; Jouault et al. 2013; Nguyen et al. 2018; Popov et al. 2020; Robertson and Roland 2008; Senses et al. 2019; Tsvetkov 1989). Indeed, the published studies reported an increase, decrease, or no change in T_g upon the addition of silica, carbon, and other reinforcing fillers (Robertson and Roland 2008). Different authors found the anomalous relaxation behavior of the interfacial polymer layer (Baeza et al. 2016; Jouault et al. 2013), some "suppression" of dynamics in this layer (Cheng et al. 2017; Popov et al. 2020), and the mobility gradient of polymer chains near a solid interface (Berriot et al. 2002; Nguyen et al. 2018). A few published numerical estimates (Baeza et al. 2016; Cheng et al. 2017; Jouault et al. 2013; Popov et al. 2020) were of importance as the prerequisites for our analysis. For example, a significant drop in mobility in the bound polymer layer of thickness $\delta = 2$–10 nm adsorbed at nanoparticles and its correlation with the size of the radius of gyration $R_g \sim 10$ nm of the bound chains (Baeza et al. 2016; Jouault et al. 2013), and a special behavior of the interfacial layer of 2–4 nm thickness (Cheng et al. 2017; Popov et al. 2020) were observed. Polymer mobility gradient near nanoparticles in nanocomposites with an increase in the relaxation time at length scales on the order of polymer radius of gyration ("stiffened" interfacial layer of 10–20 nm thick) was registered as well (Senses et al. 2019). One could suppose that polymer dynamics is likely to change the most in hybrid composites with covalently bound constituents when interparticle distances are commensurable to one or two Kuhn segments in length.

Below we present the results of calculations of the average distances d (nm) between SiO_2 units and the interfacial areas S (m^2) between covalently bound inorganic inclusions and matrix in the composite cube with one cm edge (1 cm^3) as the functions of SiO_2 content (0.03–10 wt.%) and unit size (diameter D) in the composites. For the numerical simulations of the regularities of distribution of nanoparticles or subnanounits within the CER matrix, we accepted the following assumptions. The spherical form of silica units, SiO_2 or ECH-POSS, was assumed. The boundary conditions were not applied in the calculations due to a very large number of such particles. The mass contents of silica particles were considered. The matrix mass was taken as constant at different silica contents.

First, the quantities of introduced silica units N (in 1 cm^3 of composite) of the diameter D were determined. In these calculations, the following density values were taken: 1.2 г/см^3 for the CER matrix; 2.1 g/cm^3 for SiO_2 units (as for amorphous silica), and 1.2 г/см^3 for epoxycyclohexyl (ECH)-POSS nanoparticles (1,2 + _ 0.04 g/cm^3, see POSS Handbook-Phantom Plastics, 2013, p. 27). After the calculations of N values in the cube with one cm edge (1 cm^3), the S and d values were determined.

We used Eqs. (4.1–4.6) in our calculations. On the one hand, the volume of all spherical nano- or subnanoparticles in 1 cm^3 of matrix $V_{NP} = NV_P$ where the volume of one spherical nanoparticle $V_P = \pi D^3/6$. On the other hand, $V_{NP} = m_{NP}/\rho_p$ where m_{NP} is mass of all particles in 1 cm^3 of the matrix, and ρ_p is their density. Moreover, $m_{NP} = Mc$ where M is the mass of 1 cm^3 of matrix and c is the mass content (%) of nanoparticles in this volume. As a result, we had Eqs. (4.1) and (4.2) for calculating V_{NP}, and the final formula (4.3) for N determination when we varied c and D values, whereas 6 M/$\pi\rho_P$ was the constant coefficient for the certain type of analyzed material.

$$V_{NP} = N\pi D^3/6 \tag{4.1}$$

$$V_{NP} = Mc/\rho_p \tag{4.2}$$

$$N = \frac{c}{D^3} \cdot \frac{6M}{\pi \rho_p} \tag{4.3}$$

After the calculations of N values in the cube with one cm edge (1 cm^3), the determination of the analyzed characteristics, the common interfacial areas S (between covalently bound matrix and all silica units) in 1 cm^3 of composite, and the average face-to-face distances d between embedded inorganic units were determined using the Eqs. (4.4), (4.5), and 4.6, where d_c is the distance between the centers of nanoparticles.

$$\mathrm{S} = \pi N D^2 (\mathrm{m}^2) \tag{4.4}$$

$$d_c = \frac{0.01}{\sqrt[3]{N}} (\mathrm{m}) \tag{4.5}$$

$$d = d_c - D(\mathrm{nm}) \tag{4.6}$$

Along with plotting the commonly calculated dependencies, the relevant estimates were made for the compositions of nano- and subnanocomposites studied in our works (Bershtein et al. 2016a, 2015), proceeding from the real sizes of SiO_2 units or SiO_2 cores of ECH-POSS units (Table 4.1). Since the ECH-POSS molecule includes in fact SiO_2 inner core of ca.1 nm size and the grafted outer organic shell, the real SiO_2 weight content equaled in this case to about 1/3 of the introduced ECH-POSS content. It means that the contents of introduced ECH-POSS nanoparticles of 0.03, 0.1, 0.5,

1, 2, and 10 wt.% corresponded to SiO_2 units contents in these nanocomposites of ca. 0.01, 0.03, 0.17, 0.3, 0.7, and 3.3 wt.%.

Figures 4.12 and 4.13 present the results of calculations of the values of distances *d* and interfacial areas *S* as the functions of SiO_2 content for different diameters of SiO_2 units. In these calculations, the SiO_2 unit diameter varied from 0.5 to 1.0 nm (minimal possible value of SiO_2 node in the network of subnanocomposites (Bershtein et al. 2016a)) and 1.0 nm (SiO_2 core in ECH-POSS molecule) up to 400 nm (maximal registered value of nanocluster aggregate in the nanocomposite with 10 wt.% SiO_2 (Bershtein et al. 2016a)). Figures 4.12 and 4.13 illustrate the common rule: increasing interfacial area *S* and decreasing interparticle distance *d* with increasing silica content in case if the SiO_2 unit size remains invariable. Importantly, the minimal *d* values and maximal *S* values must be expected at SiO_2 units diameters of 0.5–1 nm, i.e., in the SiO_2-containing subnanocomposites or at perfect molecular dispersion (without nanoclusterization) of the ECH-POSS nanoparticles in the nanocomposites.

Table 4.1 Sizes of inorganic particles in CER-SiO_2 and CER/ECH-POSS composites (TEM/EDXS analysis (Bershtein et al. 2016a, 2015))

CER-SiO_2 composites		CER/ECH-POSS composites	
SiO_2, wt.%	Diameter *D* of SiO_2 unit, nm	ECH-POSS, wt.%	Diameter *D* of SiO_2 core, nm
0.03	0.5–1.0*	0.03	1.0
0.1	0.5–1.0*	0.1	1.0
0.5	10	0.5	1.0
1.0	25	1.0	1.0
2.0	100	2.0	2–3**
10	400	10	70–100**

* The size of subnanometric SiO2 node in the matrix network
** Diameter of ECH-POSS nanoclusters or their aggregates

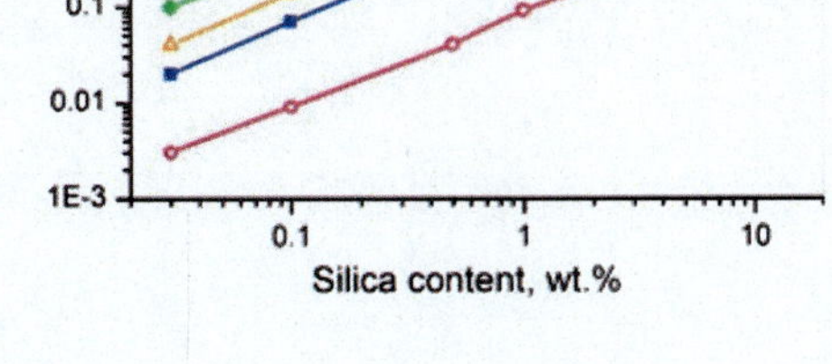

Fig. 4.12 Calculated values of the interfacial area *S* in one cm^3 of the CER-SiO_2 composites as a function of SiO_2 content and units diameter *D* (nm, indicated on the right) in the composites (Bershtein et al. 2021b)

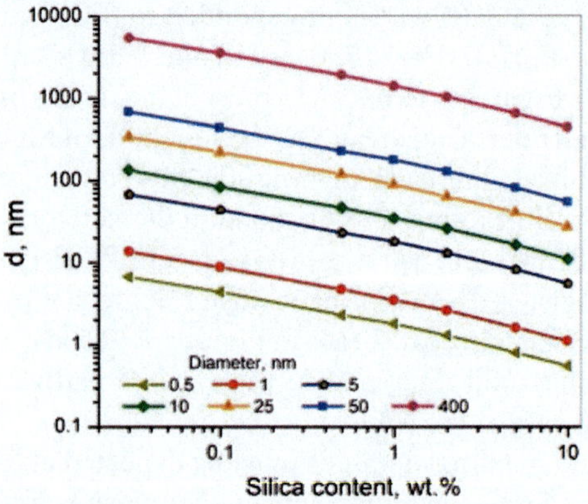

Fig. 4.13 Calculated values of the average distance d between SiO_2 units in the CER matrix as a function of SiO_2 content and unit diameter in the composites (Bershtein et al. 2021b)

Figure 4.14 presents the distance d versus silica unit diameter D dependencies calculated for the composites with SiO_2 contents varying from 0.03 wt.% to 10 wt.% (see Table 4.1). Three indicated zones of d values may be conditionally distinguished herein: zone I with minimal face-to-face distance d = 5–40 nm between SiO_2 units; zone II of increased distance d = 40–100 nm, and zone III of maximal values d = 100–500 nm.

It is logical to suppose that just the interfacial layer within zone 1 must exhibit the most "suppressed" (by SiO_2) dynamics due to the fact that such distances are commensurable with the Kuhn segments in heterocyclic macromolecules (see below). Meanwhile, dynamics in the matrix nanovolumes localized in the regions of the elevated interparticle distances (zone II in Fig. 4.14) is apparently subjected to a different influence of grafted SiO_2 units. The impact of nanoparticles on matrix

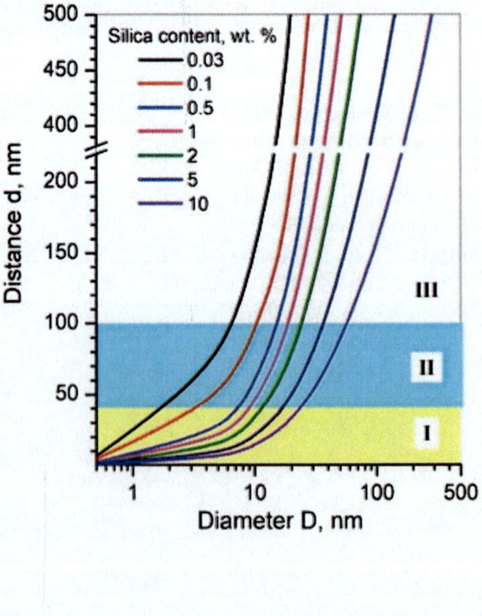

Fig. 4.14 Calculated values of the average distance d between SiO_2 units in the CER matrix as a function of SiO_2 unit diameter D in the composites with different SiO_2 contents (Bershtein et al. 2021b). The indicated zones I, II, and III are discussed in the text.

dynamics likely becomes negligibly small in the case of large face-to-face distances corresponding to zone III.

Figures 4.15 and 4.16 present *d* and *S* values as the functions of SiO_2 or ECH-POSS contents calculated proceeding from the experimentally estimated sizes of these units in composites (Table 4.1). Figure 4.15 shows that at ultra-low SiO_2 contents of 0.03–0.1 wt.% in the CER-SiO_2 composites, when nanoclusterrization was absent and only subnanometric inorganic nodes covalently embedded into the amorphous matrix could be formed (Bershtein et al. 2016a), the minimal distances $d = 6$–10 nm and maximal value of interfacial area $S = 4$ m^2 in one cm^3 of the composite are predicted. The histograms of Si distribution in nanovolumes of these composites indicated the quasi-regular distribution of SiO_2 nodes in the amorphous matrix network (Fig. 4.6). Therefore, the obtained average *d* values correspond, obviously, to the real ones, and all atomic groups of the matrix were located at distances not more than 3–5 nm from the points of attaching matrix to the SiO_2 nodes.

Unlike that, at 0.5–1 wt.% SiO_2 the distances *d* increased up to ca. 100 nm, and $d =$ 480 nm at 10 wt.% SiO_2 because of the formation of nanocluster aggregates of 400 nm

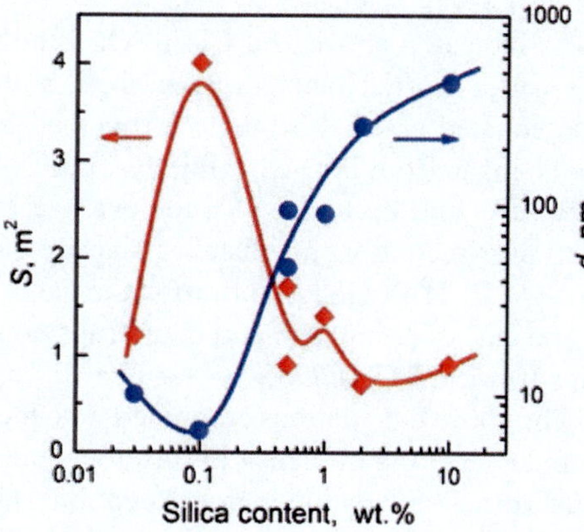

Fig. 4.15 Results of calculations of the interfacial area *S* in 1 cm^3 of the composite and the average distance *d* between SiO_2 units for the CER-SiO_2 composites with different SiO_2 contents, proceeding from the experimental estimates of SiO_2 units sizes (Table 4.1) (Bershtein et al. 2021b)

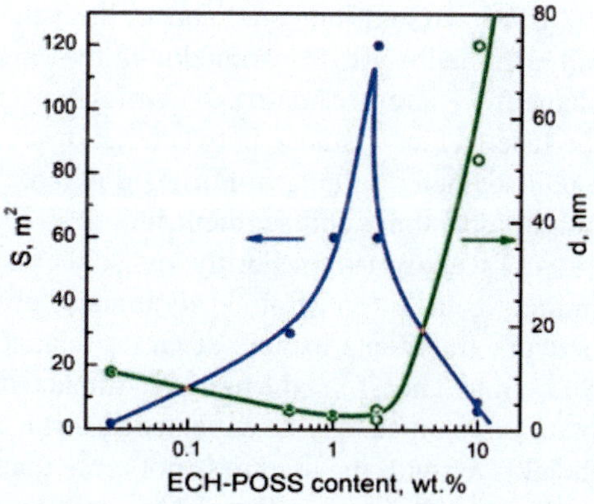

Fig. 4.16 Results of calculations of the interfacial area *S* in 1 cm^3 of the CER/ECH-POSS nanocomposites and the average distance *d* between POSS units as the functions of ECH-POSS content, proceeding from the experimental estimates of POSS units sizes (Table 4.1) (Bershtein et al. 2021b)

size (Table 4.1). Simultaneously, the values of interfacial area S dropped down to ca. 1 m^2 in one cm^3 of the composite in the latter case. Moreover, it is necessary to take into account that at 1, 2, and 10% SiO_2 structural nano- and submicroheterogeneities (nanoclusters and nanocluster aggregates) arose in the composites, and irregular distribution of SiO_2 units in the matrix was registered in the TEM/EDXS analysis. This could not only result in increasing *d* values but also lead to the wide dispersion of the interparticle distances and, therefore, to the deterioration of the properties of composites.

Figure 4.16 presents the results of similar calculations performed for the CER/ECH-POSS nanocomposites also proceeding from the real structural data (Bershtein et al. 2015). Nanoclusters could not be revealed by TEM at ECH-POSS contents of 0.03, 0.1, 0.5, and 1 wt.% when the molecular dispersion of ECH-POSS nanoparticles, obviously, prevailed although the EDXS histograms of Si nanodistribution in these composites did not strictly indicate their real quasi-regular distribution in the matrix. Some tendency to the onset of clusterization could be discerned only starting from 2 wt.% ECH-POSS in the composite, and the clusters/cluster aggregates of 70 to 100 nm size were distinctly observed when introducing 10 wt.% ECH-POSS (Table 4.1, Fig. 4.4).

As indicated above, the ECH-POSS molecule includes a silica core of 1 nm in size and a grafted outer organic shell, and the real SiO_2 weight contents in this case equaled about 1/3 of the introduced ECH-POSS content. Three conclusions can be made from Fig. 4.16. Firstly, at ultra-low ECH-POSS contents, starting from 0.03 wt.% and up to 2 wt.%, the average interparticle distance *d* does not exceed 5–20 nm. Secondly, this distance sharply increases, up to 52–74 nm at introducing 10 wt.% ECH-POSS. And thirdly, the value of interfacial area *S* may attain 100 m^2 in one cm^3 of composite and then drops down to ca. 3–5 m^2 for the nanocomposite with 10 wt.% ECH-POSS.

The above calculations combined with the structural experimental data allowed us to understand the influence of different contents of SiO_2 and POSS additives and to offer a clear explanation for the exceptional positive impact of ultra-low SiO_2 or ECH-POSS units contents on the properties of CER-based subnano- and nanocomposites (Bershtein et al. 2021b).

As known and indicated above, the values of immobilized fragments of grafted polymer chains are determined with the value of Kuhn segment length as a "rigidity element". Therefore, matrix dynamics may change most strongly in composites when the face-to-face distance between nanoparticles becomes comparable to one or two Kuhn segments in length. For rigid heterocyclic macromolecules, such as those of polyimides, the Kuhn segment length is equal to 6–20 nm (Bershtein and Egorov 1994; Tsvetkov 1989). For the densely cross-linked heterocyclic CER network, the immobilization (constrained dynamics) effect may comprise, obviously, even the network fragments located at larger distances from the points of their tethering to SiO_2 units. Therefore, the maximal impact of silica subnano- or nanounits on polymer matrix dynamics may be expected at formation of immobilized, tightly bound interfacial layer with the thickness not more than 10–40 nm that corresponds to *d* values within zone I shown in Fig. 4.14.

The above considerations testify in favor of the following conclusion. The maximal T_g values were attained for the composites with 0.03–0.1 wt.% SiO_2 units or ECH-POSS nanoparticles. This effect is the most pronounced in the case of subnanocomposites with SiO_2 nodes, e.g., due to embedding 0.1 wt.% nodes when the temperature of glass transition onset of the composite increased from 150° to 260 °C, and T_g increased from 250° to 300 °C. That may be explained by the fact that the dynamics of the whole matrix transforms into the state of *constrained interfacial dynamics*. Thus, at quasi-regular SiO_2 units distribution in the amorphous matrix (found for the subnanocomposites prepared via a sol–gel procedure), their covalent grafting to the network and contents of 0.03–0.1 wt.% the face-to-face distances between SiO_2 units equaled to 6–10 nm, i.e., were commensurable with the Kuhn segments lengths in heterocyclic macromolecules.

With increasing content of these additives, nanoclusterization and increasing sizes of nanoparticles, the interparticle distances *d* sharply increase and therefore the positive influence of nanofillers gradually decreases (zone II) and even totally disappears (zone III, Fig. 4.14). Moreover, at large content of silica particles, the properties of composite may become worse than for neat matrix (due to the sharp increasing the distances *d,* their large dispersion, and increasing the degree of network defectiveness). These considerations explain the ambivalent influence of silica units on CER dynamics and properties and the exceptional positive impact of their ultra-low contents.

Thus, the origin of the maximal impact of ultra-low SiO_2 or POSS additives contents consists in manifesting the controlling role of the CER-constrained interfacial dynamics peculiar to the whole matrix.

References

G.P. Baeza, C. Dessi, S. Costanzo, D. Zhao, S. Gong, A. Alegria, R.H. Colby, M. Rubinstein, D. Vlassopoulos, S.K. Kumar, Network dynamics in nanofilled polymers. Nat. Commun. **7**, 11368 (2016). https://doi.org/10.1038/ncomms11368

J. Berriot, H. Montes, F. Lequeux, D. Long, Evidence for the shift of the glass transition near the particles in silica-filled elastomers. Macromolecules **35**, 9756 (2002). https://doi.org/10.1021/ma0212700

V.A. Bershtein, A.M. Fainleib, L.M. Egorova, O.P. Grigoryeva, D.A. Kirilenko, S.G. Konnikov, V.A. Ryzhov, O.M. Starostenko, P.N. Yakushev, M.A. Yagovkina, J.M. Saiter, The impact of ultra-low amounts of introduced reactive POSS nanoparticles on structure, dynamics and properties of densely crosslinked cyanate ester resins. Europ. Polym. J. **67**, 128 (2015). https://doi.org/10.1016/j.eurpolymj.2015.03.022

V.A. Bershtein, A.M. Fainleib, K.G. Gusakova, D.A. Kirilenko, P.N. Yakushev, L.M. Egorova. N. Lavrenyuk, V.A. Ryzhov, Silica subnanometer-sized nodes, nanoclusters and aggregates in cyanate ester resin based networks: structure and properties of subnano- and nanocomposites. Europ. Polym. J. **85**, 375 (2016a). https://doi.org/10.1016/j.eurpolymj.2016.10.047

V.A. Bershtein, A.M. Fainleib, D.A. Kirilenko, P.N. Yakushev, K.G. Gusakova, N. Lavrenyuk, V.A. Ryzhov, Dynamics and properties of high performance amorphous cyanate ester-based subnanocomposites with ultralow silica content and quasi-regular structure. Polymer **103**, 36 (2016b). https://doi.org/10.1016/j.polymer.2016.09.020

V.A. Bershtein, A.M. Fainleib, D.A. Kirilenko, P.N. Yakushev, K.G. Gusakova, N. Lavrenyuk, V.A. Ryzhov. Incorporating silica into cyanate ester-based network by sol-gel method: structure and properties of subnano- and nanocomposites, in *Proceedings of Times of Polymers and Composites AIP TOP-2016 Conference* (Italy, June 2016c). https://doi.org/10.1063/1.4949620

V.A. Bershtein, A.M. Fainleib, P.N. Yakushev, L.M. Egorova, O.P. Grigoryeva, V.A. Ryzhov, O.M. Starostenko, Thermostable cyanate ester resins and POSS-containing nanocomposites: influence of matrix chemical structure on their properties. Polym. Adv. Technol. **27**, 339 (2016d). https://doi.org/10.1002/pat.3645

V.A. Bershtein, A.M. Fainleib, D.A. Kirilenko, P.N. Yakushev, Cyanate ester resin/Silica subnanocomposites and their superiority over nanocomposites due to fundamental role of constrained interfacial dynamics. J. Phys: Conf. Ser. **2103**, 012100 (2021a). https://doi.org/10.1088/1742-6596/2103/1/012100

V.A. Bershtein, O.P. Grigoryeva, P.N. Yakushev, A.M. Fainleib, The nature of the exceptional impact of ultra-low silica contents on the properties of cyanate ester resin based nano- and subnanocomposites. Polym. Compos. **42**, 6777 (2021b). https://doi.org/10.1002/pc.26338

V.A. Bershtein, V.M. Egorov, *Differential Scanning Calorimetry of Polymers. Physics, Chemistry, Analysis, Technology* (Ellis Horwood, New York, 1994), p. 272

V.A. Bershtein, V.A. Ryzhov, Far-infrared spectroscopy of polymers, in *Advances in Polymer Science*, vol. 114 (Springer, Berlin, Heidelberg, 1994), pp. 43–121. https://doi.org/10.1007/BFb0008694

V.A. Bershtein, A.M. Fainleib, P.N. Yakushev, D.A. Kirilenko, L.M. Egorova, O.P. Grigoryeva, V.A. Ryzhov, O.M. Starostenko, High-performance multi-functional cyanate ester oligomer-based network and epoxy-POSS containing nanocomposites: structure, dynamics, and properties. Polym. Compos. **41**, 1900 (2020). https://doi.org/10.1002/pc.25506

S. Cheng, B. Carroll, W. Lu, F. Fan, J.-M.Y. Carrillo, H. Martin, A.P. Holt, N.-G. Kang, V. Bocharova, J.W. Mays, B.G. Sumpter, M. Dadmun, A.P. Sokolov, Interfacial properties of polymer nanocomposites: role of chain rigidity and dynamic heterogeneity length scale. Macromolecules **50**(6), 2397 (2017). https://doi.org/10.1021/acs.macromol.6b02816

S. Devaraju, M.R. Vengatesan, M. Selvi, A. Ashok Kumar, I. Hamerton, J.S. Godand, M. Alagar, Low surface free energy cyanate ester–silica hybrid (CE–SiO_2) nanomaterials for low k dielectric applications. RSC Adv **3**, 12915 (2013). https://doi.org/10.1039/C3RA41144C

S. Jothibasu, A. Ashok Kumar, M. Alagar, A facile synthesis of cyanate ester silica hybrid nanocomposites by in situ sol-gel method. High Perform. Polym. **23**, 11 (2011). https://doi.org/10.1177/0954008310389838

N. Jouault, J.F. Moll, D. Meng, K. Windsor, S. Ramcharan, C. Kearney, S.K. Kumar, Bound layer in nanocomposites. ACS Macro. Lett. **2**(5), 371 (2013). https://doi.org/10.1021/mz300646a

H.K. Nguyen, X. Liang, M. Ito, K. Nakajima, Direct mapping of nanoscale viscoelastic dynamics at nanofiller/polymer interfaces. Macromolecules **51**(15), 6085 (2018). https://doi.org/10.1021/acs.macromol.8b01185

I. Popov, B. Carroll, V. Bocharova, A.-C. Genix, S. Cheng, A. Khamzin, A. Kisliuk, A.P. Sokolov, Strong reduction in amplitude of the interfacial segmental dynamics in polymer nanocomposites. Macromolecules **53**(10), 4126 (2020). https://doi.org/10.1021/acs.macromol.0c00496

C.G. Robertson, C.M. Roland, Glass transition and interfacial segmental dynamics in polymer-particle composites. Rubber Chem. Technol. **81**, 506 (2008). http://refhub.elsevier.com/B978-0-12-803581-8.02163-9/sbref21

E. Senses, S. Narayanan, A. Faraone, Nanoscale particle motion reveals polymer mobility gradient in nanocomposites. ACS Macro. Lett. **8**(5), 558 (2019). https://doi.org/10.1021/acsmacrolett.9b00176

O.M. Starostenko, V.A. Bershtein, A.M. Fainleib, L.M. Egorova, O.P. Grigoryeva, A.B. Sinani, P.N. Yakushev, Thermostable polycyanurate-polyhedral oligomeric silsesquioxane (POSS) hybrid networks: synthesis, dynamics and thermal behavior, in *Molecular Mobility and Order in Polymer Systems, Macromolecular Symposia*, ed. by K. Severing, S. Spiegel, vol. 316 (Wiley, Ltd, 2012), p. 90. https://doi.org/10.1002/masy.201290018

V.N. Tsvetkov, *Rigid-Chain Polymers. Hydrodynamic and Optical Properties in Solution* (Springer US, 1989), p. 490

Chapter 5
Other CER-Based Nanocomposites

Abstract This chapter considers the structure and properties of CER-based nanocomposites containing mesoporous silica particles, graphene oxide (GO), carbon nanotubes (CNTs), or "unzipped" carbon nanotubes (uCNTs). It is shown that by introducing glycidyl silane functionalized mesoporous silica FMCM-41, the thermal stability increases, whereas the dielectric constant and the dielectric loss tangent of the material decrease from 3.2 to 1.98 and from 0.023 to 0.009, respectively. GO produced a strong catalytic effect on the matrix curing process, and the addition of 1% GO significantly improved its mechanical properties. Of particular interest are the results of studying nanocomposites with functionalized uCNTs. In addition to significantly improving the mechanical and thermal properties of the matrix, uCNTs significantly enhanced its luminescence properties and absorption in the UV region.

5.1 Nanocomposites with Mesoporous Silica Particles: Materials with Low Dielectric Constant

The development of microelectronics requires materials with low dielectric constant *k*. Generally, microelectronic integrated circuits consist of interconnected wires which are insulated by dielectric materials. In order to reduce the capacitance, it is essential to replace the present insulating materials for thin low-*k* dielectric films. There are two main ways commonly used to reduce the dielectric constant of insulating materials: (1) by the introduction of less polar atomic groups into the polymer matrix skeleton, and (2) by the preparation of polymer nanocomposites with porous silica materials.

Ariraman et al. (2014) have developed a new type of cyclohexyl branched aliphatic chain bridged cyanate ester. First, the 1,3-bis(4-cyanatobenzyl) cyclohexane (BCC) monomer was synthesized. Cyanate ester nanocomposites have been developed by polymerizing this cyanate ester monomer reinforced with varying weight percentages of glycidyl silane functionalized mesoporous silica particles MCM-41 (FMCM-41). The BCC-based CER/FMCM-41 nanocomposites with different concentrations of

V. A. Bershtein and P. N. Yakushev, *High-Temperature Polymer Nanocomposites Based on Heterocyclic Networks from Nitrile Monomers*, Springer Series in Materials Science 334, https://doi.org/10.1007/978-3-031-32943-2_5

FMCM-41 (1, 3, 5, 7, and 10 wt.%) were prepared. The curing process was performed by heating up to 210 °C with the rate of 20 °C/hr. Figure 5.1 shows schematically the route to obtain BCC-based CER/FMCM-41 nanocomposites.

In (Ariraman et al. 2014), the monomer and polymer composites were characterized by spectral analysis using FTIR, ^{1}H and ^{13}C NMR spectroscopy. The thermal properties were analyzed by TGA and DSC, and the morphological state was studied by SEM and high-resolution transmission electron microscopy (HRTEM). The disappearance of absorption bands at 2252 cm^{-1} (the –OCN group of the cyanate ester) with the appearance of absorption bands at 1085 cm^{-1} and 1365 cm^{-1} was associated with the successful formation of triazine rings by the polymerization of OCN groups and with the formation of Si–O–Si linkages. Moreover, the simultaneous disappearance of absorption bands at 2252 cm^{-1} and 943 cm^{-1} (the glycidyl group of FMCM-41), and the appearance of the band at 1645 cm^{-1} (oxazoline ring) indicated the successful incorporation of FMCM-41 through covalent bonding by cyclic oxazoline ring formation between the OCN group of the cyanate ester and the glycidyl group of FMCM-41. The structural HRTEM/SEM analysis clearly indicated the

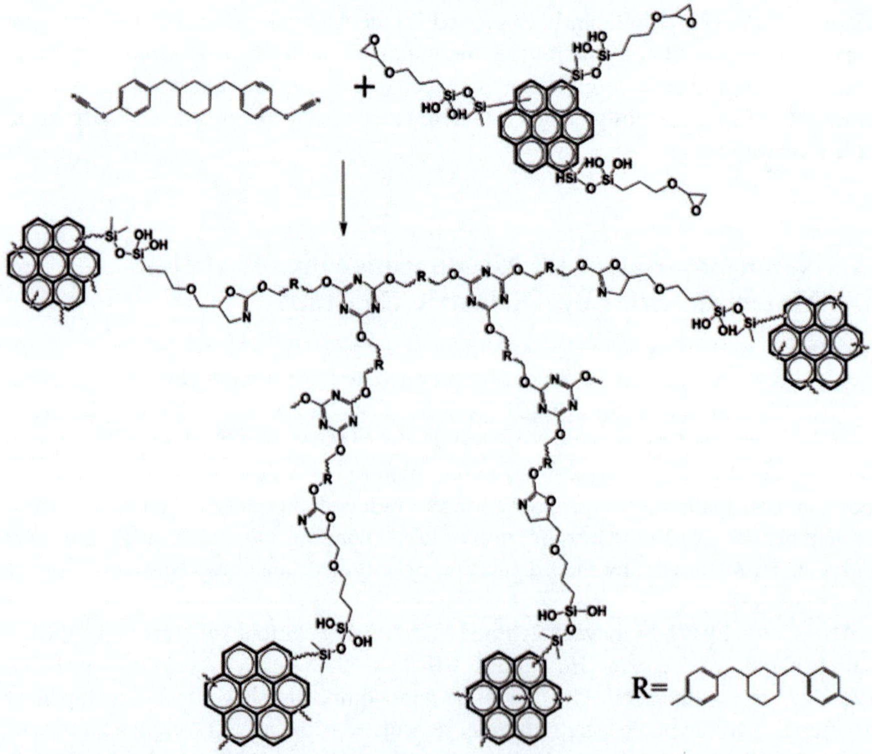

Fig. 5.1 Scheme of the BCC-based CER/FMCM-41 polymeric nanocomposite (Ariraman et al. 2014). BCC is 1,3-bis(4-cyanatobenzyl) cyclohexane. FMCM-41 is glycidyl silane functionalized mesoporous silica material MCM-41

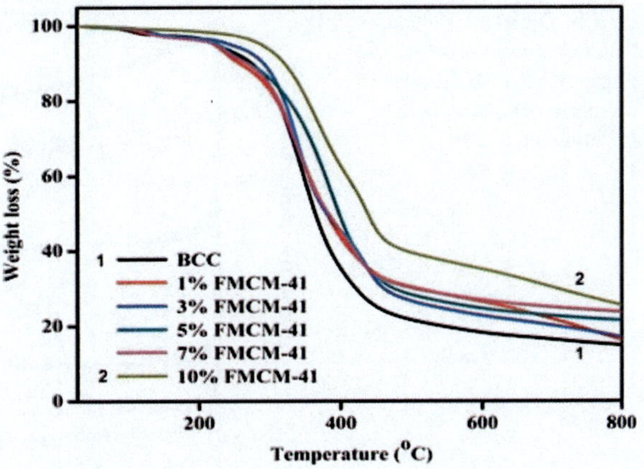

Fig. 5.2 TGA curves of neat BCC-based matrix and the BCC-based CER/FMCM-41 polymeric nanocomposites (Ariraman et al. 2014)

existence of pores and the uniform distribution of the porous silica inclusions in the BCC-based CER/FMCM-41 composites (Ariraman et al. 2014).

It was found that increasing weight percentages of FMCM-41 in the composites enhanced their thermal stability and strongly reduced the value of the dielectric constant.

The thermal stability of the BCC-based CER/FMCM-41 composites has been ascertained by TGA. Figure 5.2 shows the TGA curves for the neat matrix and several composites. It can be seen that FMCM-41 increased the thermal stability of the resulting polymer composites, and the composite with the highest percentage of 10% mesoporous silica exhibited the better thermal stability than the neat matrix sample.

Figure 5.3 shows the dielectric constant *k* vs. frequency curves obtained for the neat BCC-based CER matrix and the series of the relevant CER/FMCM-41 nanocomposites. The neat CER matrix exhibited a dielectric constant of about 3.3 at frequency of 10^5 Hz and 3.0 at 1 MHz; these values are lower than that for the previously reported polycyanurates. Thus, the existence of less polar groups in the cyanate ester monomer skeleton significantly reduced the value of the dielectric constant in the neat polycyanurate; however, it was not sufficient to meet the low *k* requirements.

At the same time, there is a pronounced tendency for decreasing the dielectric constant as a consequence of the introduction of mesoporous functionalized FMCM-41 particles, as well as with increasing their content in the composite. This is primarily due to an increase in the content of air-filled voids in the composites. Thus, the lowest dielectric constant of 1.98 at 1 MHz was obtained for BCC-based CER composites with 10 wt.% FMCM-41 particles uniformly distributed in the matrix.

In addition, as the content of mesoporous FMCM-41 particles in these composites increased, the value of dielectric losses significantly decreased. As can be seen from the dielectric loss spectra given in Ariraman et al. (2014), the dielectric loss tangent value was of 0.0233 for the neat matrix, and its minimum value of 0.0092 was registered for the composite with 7 wt.% FMCM-41 (Ariraman et al. 2014). These experimental data suggest that this newly developed class of thermally stable

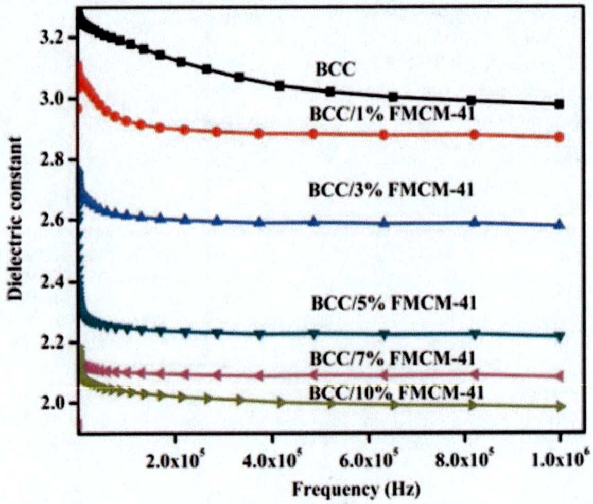

Fig. 5.3 Dielectric constants of BCC-based matrix and the relevant CER/FMCM-41 polymeric nanocomposites (Ariraman et al. 2014)

CER polymer nanocomposites can be used as a potential insulation material for microelectronic applications.

Here it is also relevant to mention the work (Sun et al. 2013), in which multifunctional composites based on DCBE-based CER matrix and 10% and 20% SiO_2 coated Fe_3O_4 submicrometer-scale spherical particles (a conducting core/insulating shell configuration) were obtained by hydrothermal method and studied. The dielectric constant of the resulting composites was found to be enhanced over a wide frequency and temperature ranges, while the low dielectric loss tangent of the neat CER was largely preserved up to 160 °C due to the insulating SiO_2 coating on individual conductive Fe_3O_4 spheres. These composites also demonstrated high dielectric breakdown strengths at room temperature. DMA indicated that the storage modulus of the composite with a 20 vol.% filler loading was twice as high as that of the neat matrix. Magnetic measurements revealed a large saturation magnetization and negligibly low coercivity and remanent magnetization in these composites (Sun et al. 2013).

5.2 Nanocomposites with Carbon Nanofillers (Graphene, Nanotubes)

Graphene as an atomically thick, two-dimensional (2-D) substance, and one-dimensional (1-D) carbon nanotubes (CNTs) with outstanding mechanical, thermal, and other properties are the most known among the carbon nanofillers. In order to develop high-performance polymer nanocomposites with these carbon nanofillers, two basic problems should be solved: (1) achieving good dispersion of the nanofiller in the polymer matrix and (2) providing improved adhesion and enhanced interaction at the boundaries between the matrix and the carbon nanofiller. Therefore, the functionalization of carbon nanofillers is the most common method for solving these problems. So the most typical method of functionalization of graphene (or the source material, graphite) is an oxidation process with grafting chemically of hydroxyl, carbonyl, and epoxide groups to the nanocarbon surface. This allowed to prepare a number of polymer-graphene nanocomposites (Stankovich et al. 2006; Ramanathan et al. 2008; Kim et al. 2010; Cai and Song 2010).

Last decade, several works have been performed on CER nanocomposites containing graphene (Wang et al. 2012; Lin et al. 2013a) or carbon nanotubes (Fainleib et al. 2012; Lin et al. 2013b; Rakesh et al. 2015; Lu et al. 2018; Yang et al. 2021; Hu et al. 2022; Li et al. 2018). Wang et al. (Wang et al. 2012) studied the curing dynamics and network formation of PT-30-based CER/graphene oxide (GO) nanocomposites, whereas Lin et al. (2013a) prepared and studied properties of DCBA-based CER/graphene oxide nanosheets (GONSs) composites. Graphene oxide was synthesized from expandable graphite.

The curing dynamics and network formation of PT-30-based CER/GO nanocomposites were studied in Wang et al. (2012) by means of DSC, FTIR, and Raman spectroscopy. FTIR and Raman spectra were utilized to reveal the chemical reactions that happened in the curing process. It was confirmed that (a) the GO was functionalized with hydroxyl, carboxyl, and epoxide groups; (b) these functional groups reacted with cyanate groups of the forming matrix; and (c) the triazine cycles were formed in the matrix network (Wang et al. 2012).

In (Wang et al. 2012), two main effects due to the influence of GO on the PT-30-based CER matrix were found. First, the GO additive produced a strong catalytic effect on the cyclotrimerization reaction (creation of triazine cycles) of the PT-30 oligomer. Figure 5.4 shows the non-isothermal DSC plots for curing of the PT-30/GO systems.

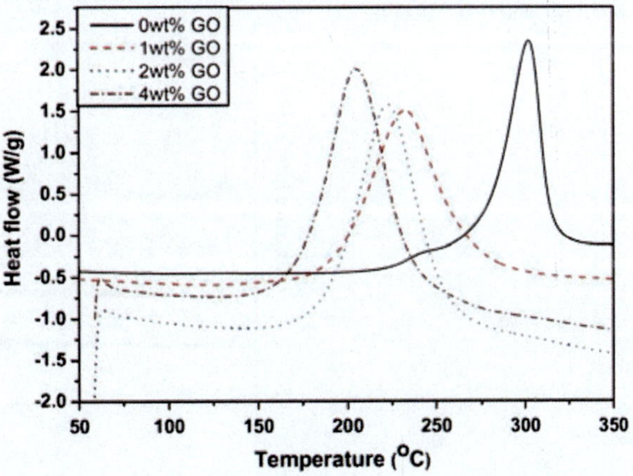

Fig. 5.4 Non-isothermal DSC plots (exotherms) for curing of the PT-30/GO systems (Wang et al. 2012)

One can see that the onset and peak temperatures dramatically decreased with incorporating GO and increasing its content from 1 wt.% to 4 wt.%. For the blend with 4 wt.% GO, the peak temperature decreased by 97 °C compared with that for the neat PT-30 oligomer. The presence of GO also contributed to a low and constant activation energy of the curing process till its end (Wang et al. 2012). The opposite effect was induced with the further increasing GO content in the composites, and it was caused by the space hindrance due to its 2D nature. This effect consisted in elevating the activation energy to some extent with increasing CO content. SEM images showed the influence of the GO on the morphology of the nanocomposites.

Lin et al. (2013a) described the synthesis and properties of graphene oxide nanosheets (GONSs)/CER nanocomposites. In this paper, isocyanate-treated graphene oxide (iGO) and DCBA-based CER were used to prepare nanocomposites, and then their structures and properties were studied. The structures of the GONSs and the nanocomposites were studied using FTIR spectroscopy, *X*-ray diffraction (XRD), TEM, and SEM. The mechanical and tribological properties of the nanocomposites were also investigated, and their thermal behavior was characterized by TGA.

FTIR spectra confirmed the successful functionalization of GO. The XRD patterns of GO, iGO, and the composites (Fig. 5.5) showed that the diffraction peak of GO corresponded to the interlayer spacing of about 0.85 nm, whereas the interlayer spacing of 0.94 nm was determined for iGO.

In addition, the nanocomposites exhibited no reflection peaks in the XRD patterns, i.e., the iGO was effectively exfoliated into isolated graphene nanosheets within the matrix. The individual nanosheets had sizes extending from tens to several hundreds of square nanometers. TEM images of the GONSs and the composites with 1.0 wt.% GONSs showed that graphene nanosheets tended sometimes to congregate together to form multilayer agglomerates; however, the exfoliated GONSs were observed due to the ultrasonic treatment of iGO in dimethylformamide.

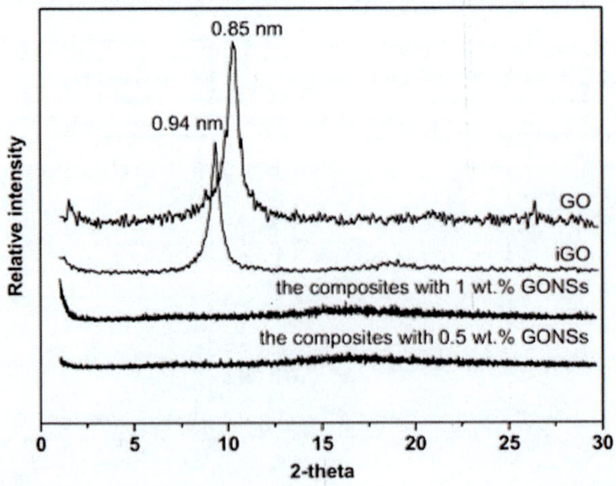

Fig. 5.5 XRD patterns of GO, iGO, and GONSs (Lin et al. 2013a)

The flexural strength, impact strength, and microhardness of the nanocomposites with various GONSs contents are shown in Fig. 5.6. It can be seen that the flexural strength and impact strength of the nanocomposites are improved essentially, up to about 1.5 times, with increasing GONSs contents up to 1.0 wt.%. This improvement in flexural strength and impact strength could be related to the excellent mechanical properties of graphene nanosheets. The edges of GONSs had functional atomic groups which could react with the prepolymer, thus providing strong interfacial bonding between the GONSs and the matrix. However, the flexural strength and impact strength of the composite with higher GONSs contents could not be improved decreasing in particular when the GONSs content was 1.25 wt.% due to the aggregation of GONSs. Representative SEM images of impact fracture surfaces of the composites with different GONSs contents showed also the typical changes in their characteristics, in particular, large separated clusters in the matrix existed due to the aggregation of GONSs for the composite with 1.25 wt.% GONSs.

Hardness can be said to represent the resistance of the material surface to abrasion, scratching, and cutting; hence, the hardness of a material has a close relationship with its tribological behavior. Figure 5.6 shows that the microhardness increased with increasing GONSs content.

The curves of variation of friction coefficient with sliding time for the composites with various GONSs contents are shown in Fig. 5.7. It can be seen that the friction coefficient in the steady-state period decreases with increasing GONSs content in the nanocomposites. Thus, the addition of GONSs is beneficial to improve the mechanical and tribological properties of the composites.

Finally, TGA showed that the composites possessed some better thermal stability in comparison with the neat CER matrix, in particular the nanocomposite with at 1.0 wt.% GONSs had higher char residue at 800 °C (45%) than that for the neat CE resin matrix (35%) (Lin et al. 2013a).

Lin et al. (2013b) demonstrated the influence of carbon nanotubes (CNTs) on the curing process and network formation of PT-30 oligomer-based Cyanate Ester (CY) resin. The effect of pristine CNTs and three functionalized CNTs on the curing process and network formation of CY was investigated by means of DSC, modulated temperature DSC, field emission gun scanning electron microscopy (FEG-SEM), FTIR, and Raman spectroscopy.

Incorporation of the various CNTs showed different accelerating effects on the curing of the CY. Figure 5.8 shows DSC plots for the CY/CNT systems at heating in a nitrogen atmosphere (60 ml/min) at the rate of 10 °C/min.

One can see that the addition of the pristine multi-walled carbon nanotubes (MWCNTs) did not exhibit a prominent accelerating effect, whilst the addition of MWCNT-OH, MWCNT-COOH, and single-walled carbon nanotubes (SWCNT)-OH showed an accelerating effect. At the limit, the onset temperature of the curing exotherm shifted from 217 °C to 170 °C with the incorporation of only 1 wt.% MWCNT-OH. With 4 wt.% MWCNT-OH, the peak temperature decreased by 57 °C. Furthermore, with the addition of CNTs, the reaction enthalpy $\Delta H_{\mathrm{curing}}$ calculated from the dynamic DSC runs, became much larger than that of neat CY increasing from 500 J/g to 640–660 J/g. Since the reaction enthalpy is directly related to the

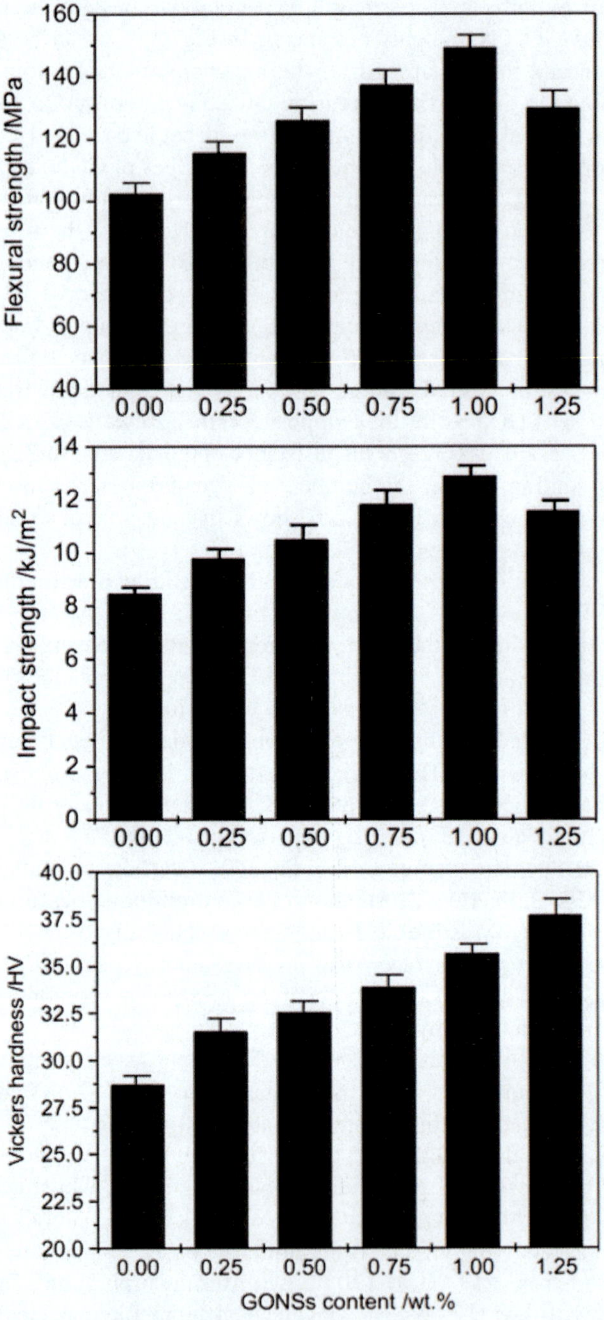

Fig. 5.6 Mechanical properties (flexural strength, impact strength, and hardness) of the nanocomposites with various GONSs contents (Lin et al. 2013a)

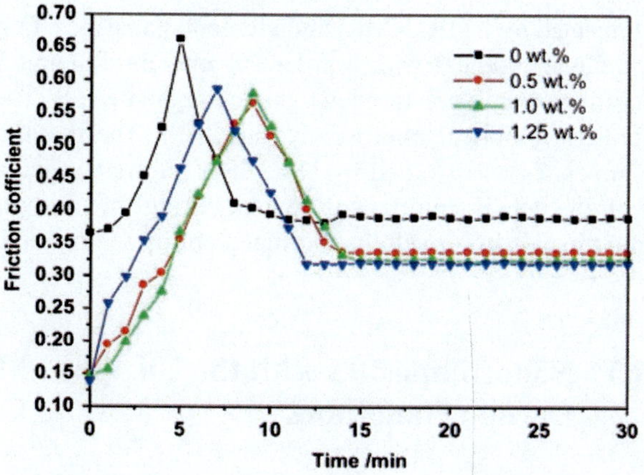

Fig. 5.7 Variation curves of friction coefficient with sliding time for the nanocomposites with various indicated GONSs contents (Lin et al. 2013a)

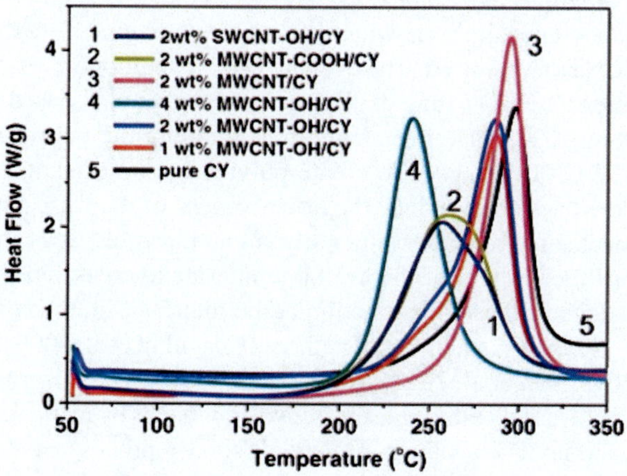

Fig. 5.8 DSC exotherms characterizing curing of the CY/CNTs mixtures at heating in nitrogen atmosphere (60 ml/min) at the rate of 10 °C/min (Lin et al. 2013b)

cross-linking degree of the CY, this substantial change in reaction enthalpy implied that the addition of the CNTs could significantly affect the network formation of the CY. The curing time of the cyanate ester also decreased with an increase in CNTs loading. The FTIR spectra monitored both the formation of triazine rings in the CY matrix and its hybridization with CNTs in the nanocomposites. Nanoscale dispersion of both the pristine and the functionalized MWCNTs in the CY matrix was observed by using FEG-SEM. In fact, all the MWCNTs were well dispersed at a nanoscale level in the CY matrix, with no indication of aggregation taking place. The up-shifting of the bands for CNTs in Raman spectra confirmed the nanoscale dispersion of MWCNTs in the CY matrix and the strong interaction between the MWCNTs and the CY.

Rakesh et al. (2015) described a novel method for the preparation of porous cyanate ester (CE) networks by using the self-assembly and organo-gelation of the monomers and subsequent curing of the xerogels. Structures of these networks were confirmed by FTIR, NMR, and elemental analyses. Organogels were prepared from the CE monomers with a polar solvent. Introducing SWCNTs into the monomer solution prior to gelation led to nanocomposite gels. The morphology of the gels was studied by optical microscopy and SEM. The polymerization of the gel/SWCNT composites was studied by DSC. The curing temperatures and time were reduced with the gels in comparison to neat curing. In addition, the incorporation of CNTs into the gels improved the thermal stability of nanocomposites.

5.3 Nanocomposites with the Unzipped Multi-walled Carbon Nanotubes

Of particular interest in the topic under consideration are the recent works of Tang, Fainleib, et al. (Yang et al. 2021; Hu et al. 2022) and Li et al. (2018) on CER-based nanocomposites, in which so-called "unzipped" carbon nanotubes (uMWCNTs, uCNTs) were used. The cyanate ester resin monomers, bisphenol A-based dicyanate ester (DCBA) (Yang et al. 2021) or bisphenol E-based dicyanate ester (DCBE) (Hu et al. 2022) have been taken for obtaining the matrices networks. Really, although MWCNTs are used to create polymer composites, they are known to have certain drawbacks that reduce the effectiveness of their use for this purpose, especially the mechanical characteristics of nanocomposites. It is their closed tubular structure, insufficient size of the available interfacial areas, smooth surface of nanotubes, and reduced adhesive interaction at the interfaces in the composites.

Recently, a number of authors (Kosynkin et al. 2009; Zhang et al. 2009; Wang et al. 2010; Fan et al. 2012; Mahmoud et al. 2015) have proposed chemical methods that could largely eliminate the above drawbacks by using methods of "unzipping" carbon nanotubes—to obtain uMWCNTs. The process of unzipping MWCNTs consists of oxidative treatment of nanotubes, leading to a longitudinal cutting and opening of their walls, making them more accessible. At the same time, depending on the conditions of this process different degrees of unzipping occur, with increasing the external radius of the MWCNTs, in the limit up to the transformation of nanotubes into graphene ribbon-like nanolayers. As a result, the contact area of the matrix with nanotubes increases dramatically. But it is also extremely important that in addition to the above treatment with "unraveling" the nanotubes, the surface of the carbon nanolayers is chemically modified, and therefore the interfacial interactions in nanocomposites are significantly enhanced. As a result, the MWCNTs possessed many oxygen-containing groups, such as –OH and –COOH groups, which reacted with the –O–C≡N groups of CE monomer and growing CER network. This provided the strong interfacial covalent bonding between uMWCNTs and the CER matrix. Additionally, the dispersion of uMWCNTS was improved due to the increase of functional groups on their surfaces.

After this brief introduction, we will turn to an analysis of the works (Yang et al. 2021; Hu et al. 2022). The main feature of these original studies is, in addition to their use of uMWCNTs in comparison with MWCNTs, the achievement in the complex of good thermal properties, improved mechanical characteristics, and imparting important functional luminescent properties to nanocomposites. The latter was achieved by introducing europium (Eu) complexes (Yang et al. 2021) or Eu-containing complexes including two ligands (ETP) (Hu et al. 2022); the latter was chosen because it produces high fluorescence and quantum yield.

Figure 5.9 shows the TEM images and the histograms of diameter distribution for carbon nanotubes in different (initial, unzipped, and ETP-anchored) states. One can see that the initial MWCNTs have hollow tubular structure and smooth surface (**a**) whereas the uMWCNTs (**b**) are characterized by a partially broken structure; the average tube diameters are increased after unzipping from 23.92 nm/43.80 nm to 38.55 nm/77.68 nm, which shows that the longitudinal width and the specific surface area are increased. The uMWCNTs structure is transformed into a graphene nanoribbon one. Figure 5.9, **c** shows that doping of ETP does not destroy the original morphology of carbon nanotubes. The smooth wall of MWCNTs became rough indicating that ETP was successfully adsorbed on the nanotubes' surfaces. Figure 5.9, **d** shows that the nanotube structure is unzipped, and the ETP is attached to the surface of the uMWCNTs.

In (Yang et al. 2021), the concentration of uMWCNTs in the CER/uMWCNT or the CER/uMWCNT/Eu nanocomposites was 0.1 wt.%. In (Hu et al. 2022), for

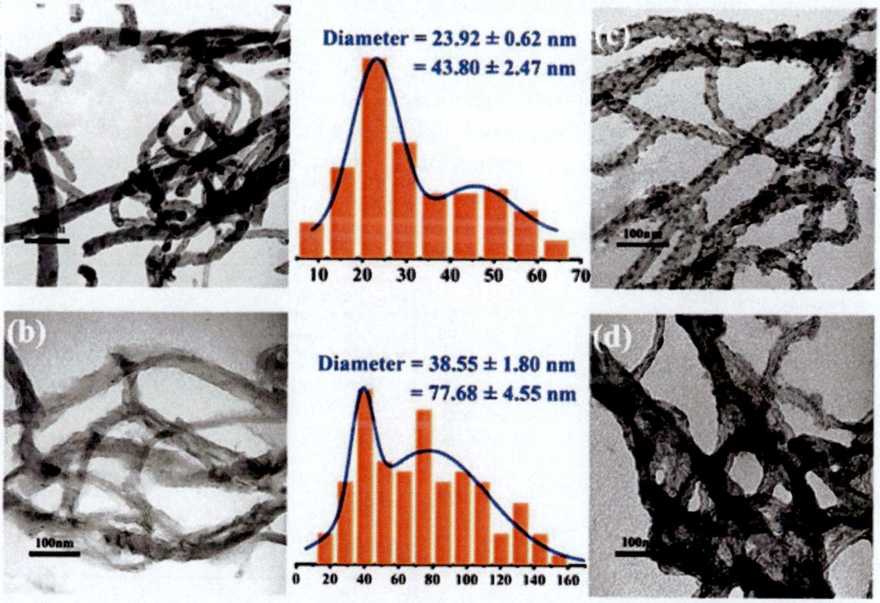

Fig. 5.9 TEM images and average diameter distribution histograms of **a** MWCNTs, **b** uMWCNTs, **c** ETP-MWCNTs, **d** ETP-uCNTs (Hu et al. 2022)

the preparation of the composites, a certain amount (0.2 wt.%, 0.4 wt.%, 0.6 wt.%, 0.8 wt.%, 1.0 wt.%) of MWCNTs, uMWCNTs, ETP-MWCNTs or ETP-uMCNTs was used. The compositions were polymerized according to the following curing schedule: 100 °C/3 h + 150 °C/2 h + 180 °C/2 h + 210 °C/1 h + 230 °C/1 h (Hu et al. 2022).

Figure 5.10 again demonstrates (as above Figs. 5.4 and 5.8) the strong catalytic effect of the introduced nanotubes and complexes on the CER curing process: the temperature of the exotherm curing maxima is reduced by ~ 40–70 °C for nanocomposites relative to the neat matrix.

Figures 5.11, 5.12, and 5.13 demonstrate a significant enhancement of the mechanical properties of the CER matrix with the introduction of the discussed additives, namely, unzipped carbon nanotubes, especially in combination with ETP and Eu complexes.

Figure 5.11 shows flexural stress–strain curves for the pure CER-E matrix and for a series of CER-E-based nanocomposites with 0.8 wt.% of the indicated additives. It can be seen that the flexural strength of neat CER-E matrix increases in the series of nanocomposites: the CER-E/MWCNTs, CER-E/uMWCNTs, CER-E/Eu complex (ETP), and CER-E/ETP-uCNTs by 14.3%, 40.8%, 59.4%, and 92.8%, respectively. Thus, the combination of two types of different additives, ETP and unzipped nanotubes, provides a greater reinforcement effect than individual MWCNTs or uMWCNTs. The maximum effect in the latter case was obviously provided by the combined influence of three factors: (a) the creation of coordination bonds between Eu3 + and O and N atoms in the molecular structure of the matrix; (b) the formation of covalent bonds between oxygen-containing groups of uMWCNTs (or uCNTs) and nitrile groups of the forming matrix network, and (c) with the improved dispersion of nanoparticles in the matrix (Hu et al. 2022). The transition from physical interactions at interfaces in nanocomposites in the case of unmodified nanotubes to chemical interactions has contributed to improved interfacial adhesion and increased the role of suppressed interfacial matrix dynamics in the formation of the better mechanical properties of nanocomposites.

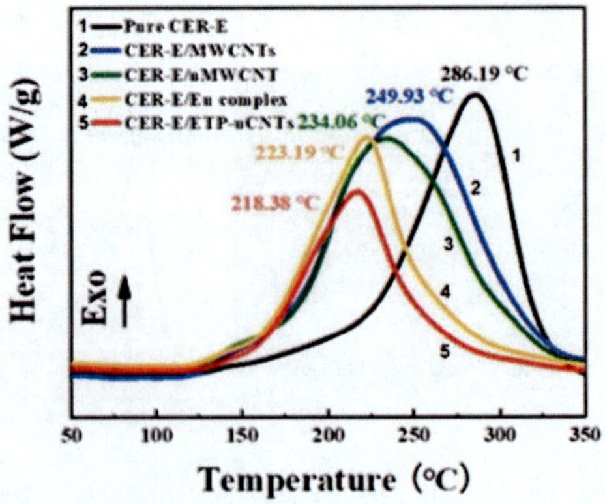

Fig. 5.10 DSC thermograms of CER-E curing process of neat CER-E matrix and CER-E hybrid nanocomposites (Hu et al. 2022)

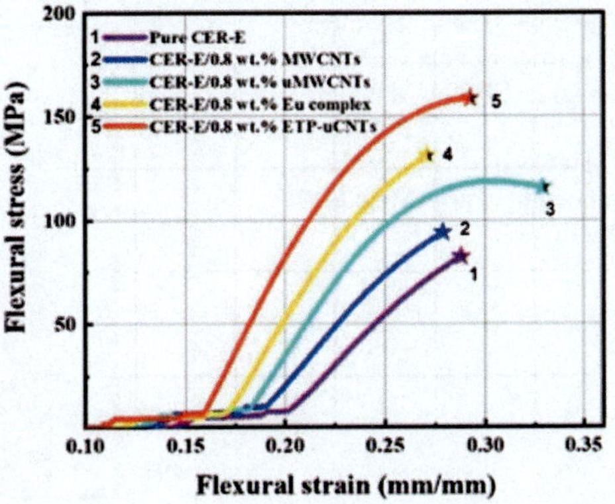

Fig. 5.11 Flexural stress–strain curves of the neat CER-E network, CER-E/MWCNTs, CER-E/uMWCNTs, CER-E/Eu complex (ETP), CER-E/ETP-uCNTs containing 0.8 wt.% of the nanofillers (Hu et al. 2022)

Figure 5.12 presents the dependencies of mechanical properties of the examined nanocomposites as a function of the concentration of ETP-uCNTs introduced into them. It can be seen that the tensile strength, flexural strength, and impact strength increase in nanocomposites, reaching maximum values 1.5–2.0 higher than in the neat matrix, at ~ 1 wt. % ETP-uCNTs. With an increase in the mass concentration of these additives, the strength of the composites did not increase, and at concentrations > 2 wt.%, the opposite trend was observed—a decrease in strength due to the aggregation of nanoparticles.

Comparison of tensile and flexural strengths of neat CER matrix and nanocomposites based on it, containing only 0.1 wt.% MWCNTs or unzipped MWCNTs are given in Fig. 5.13. The significant effect of the unzipping nanotube procedure is evident: compared to the neat matrix, the flexural strength and tensile strength of the nanocomposites are increased by 76.3% and 50.9%, respectively.

SEM images of the studied nanocomposites and their tensile fracture surfaces were also used to characterize the dispersity of MWCNTs and uMWCNTs in the matrix, the character of the fracture of composites, and their possible strengthening

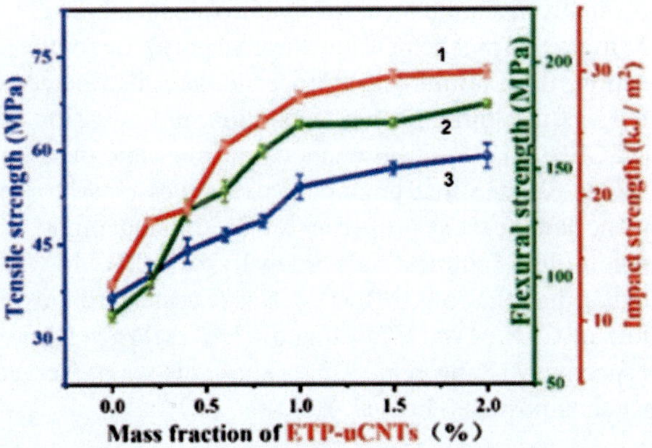

Fig. 5.12 Concentration dependencies of mechanical properties (tensile strength, flexural strength, and impact strength) for CER-E/ETP-uCNTs hybrid nanocomposites (Hu et al. 2022)

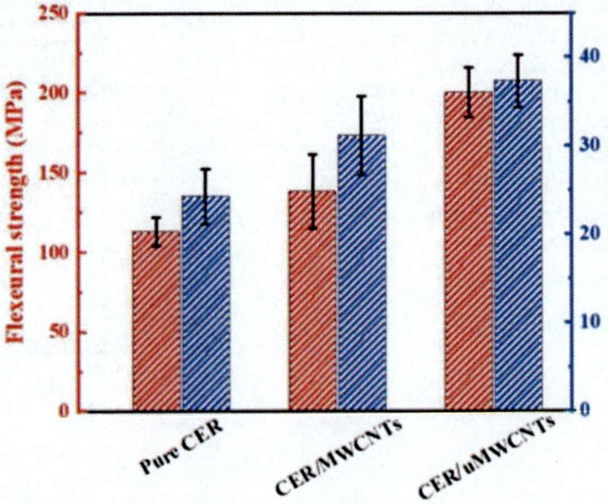

Fig. 5.13 The mechanical properties of neat CER-A network, CER-A/MWCNTs (0.1 wt.%), and CER-A/uMWCNTs (0.1 wt.%) nanocomposites (Yang et al. 2021)

mechanisms (Yang et al. 2021; Hu et al. 2022). Depending on the nanocomposite composition, the SEM was able to observe very significant differences both in the nanostructure of the investigated composites and in the nature of their fracture.

On the one hand, agglomerates of nanotubes were observed in the CER/MWCNTs nanocomposites, i.e., unsatisfactory dispersion of nanoparticles. This and the weak interactions at the interfaces naturally contributed to the occurrence and growth of microcracks. The reduced strength, in this case, was accompanied, as shown, by the fact that MWCNTs have been pulled out of the matrix when the composite fractures. The increased area at the interphase boundaries and the sharp increase in the oxygen-containing groups in uMWCNTs as a result of the unzipping procedure led to the strong interaction between nanoparticles and CER matrix and to a change in the nature of nanocomposite fracture, which occurred with enhanced toughening effect and without obvious pulling out nanotubes from the matrix. The best dispersion of nanoparticles in the matrix and enhancement of the toughness were observed in the case of composites containing ETP-uCNTs (Hu et al. 2022).

Finally, it should be noted that in Yang et al. (2021); Hu et al. 2022), in addition to the improved mechanical and thermal properties of the nanocomposites under consideration, their luminescent properties are also detected and analyzed. Figure 5.14 presents the luminescence excitation and emission spectra of CER/MWCNTs/Eu and CER/uMWCNTs/Eu nanocomposites and their luminescence decay curves.

One can see that unzipping nanotubes clearly contributed to the enhancement of the luminescent properties of these nanocomposites and absorption in the UV region. The luminescence intensity of CER/uMWCNTs/Eu nanocomposite was higher than that of CER/MWCNTs/Eu nanocomposite, and the luminescence lifetime of CER/MWCNTs/Eu and CER/uMWCNTs/Eu was 8.61 μs and 186.39 μs, respectively (Yang et al. 2021). The same was observed for the CER-E/ETP-uCNTs nanocomposites (Hu et al. 2022).

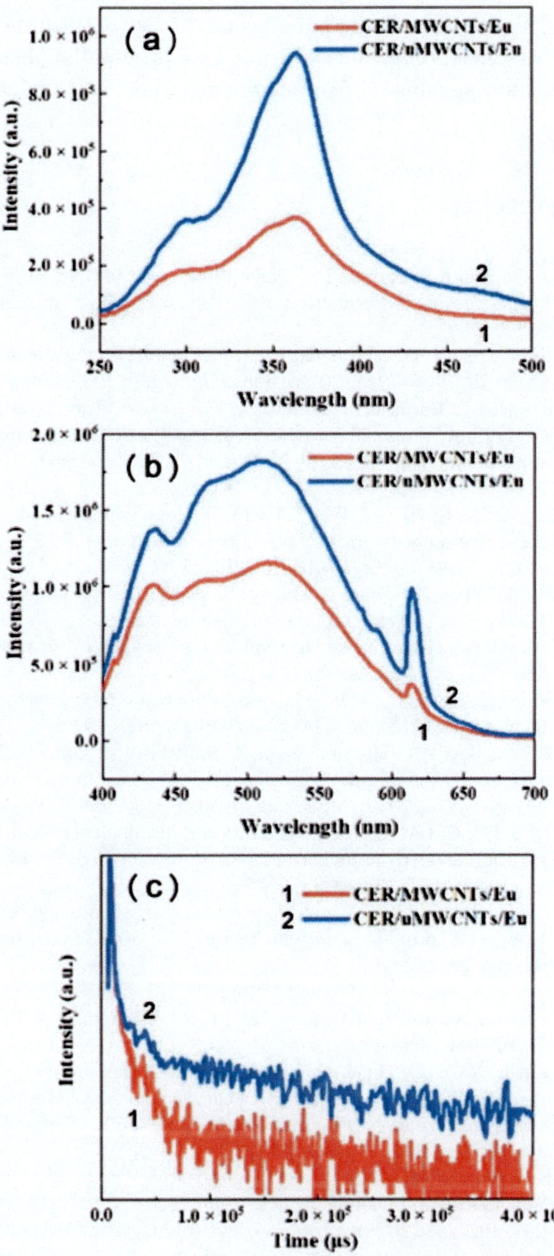

Fig. 5.14 Luminescence excitation (**a**) and emission (**b**) spectra and luminescence decay curves (c) of CER/MWCNTs/Eu and CER/uMWCNTs/Eu nanocomposites (Hu et al. 2022)

Thus, it is natural to believe that the new CER/uMWCNTs/Eu and CER-E/ETP-uCNTs nanocomposites will find also different applications including fluorescence detection, sensing, UV protective materials, etc., including high temperatures.

References

M. Ariraman, R.S. Kumar, M. Alagar, Studies on FMCM-41 reinforced Cyanate Ester nanocomposites for low k applications. RSC Adv. **4**, 57759–57767 (2014). https://doi.org/10.1039/C4RA09399B

D. Cai, M. Song, Recent advance in functionalized graphene/polymer nanocomposites. J. Mater. Chem **20**, 7906–7916 (2010). https://doi.org/10.1039/C0JM00530D

A. Fainleib, L. Bardash, G. Boiteux, O. Grigoryeva, Thermosetting cyanate ester resins filled with CNTs. In Advances in progressive thermoplastic and thermosetting polymers, perspectives and applications, vol. 10, ed. by Y. Mamunya, M. Iurzhenko, (Tehnopress editura, Iasi, Romania, 2012) p. 379

J. Fan, Z. Shi, M. Tian, J. Wang, J. Yin, Unzipped multiwalled carbon nanotube oxide/multiwalled carbon nanotube hybrids for polymer reinforcement. ACS Appl. Mater. Interfaces **4**, 5956–5965 (2012). https://doi.org/10.1021/am301623t

Z. Hu, D. Zhao, Y. Wang, L. Huang, S. Wang, S.I Mao, O. Grigoryeva, P. Strizhak, A. Fainleib, J. Tang, The created excellent thermal, mechanical and fluorescent properties by doping Eu3+-complex-anchored carbon nanotubes in polycyanate resins. Nanomaterials 12(12), 2040 (2022). https://doi.org/10.3390/nano12122040.

H. Kim, A.A. Abdala, C.W. Macosko, Graphene-polymer nanocomposites. Macromolecules **43**, 6515–6530 (2010). https://doi.org/10.1021/ma100572e

D.V. Kosynkin, A.L. Higginbotham, A. Sinitskii, J.R. Lomeda, J.R. Loeda, A. Dimiev, B.K. Price, J.M. Tour, Longitudinal unzipping of carbon nanotubes to form graphene nanoribbons. Nature **458**, 872–876 (2009). https://doi.org/10.1038/nature07872

H. Li, J. Gu, C. Liu, D. Wang, Thermal and mechanical properties of cyanate ester resin modified with acid-treated multiwalled carbon nanotubes. High Perform. Polym. **30**, 38–46 (2018). https://doi.org/10.1177/0954008316675208

Q. Lin, L. Qu, Q. Lü, C. Fang, Preparation and properties of graphene oxide nanosheets/Cyanate Ester resin composites. Polym. Testing **32**, 330–337 (2013a). https://doi.org/10.1016/j.polymertesting.2012.11.014

Y. Lin, C.A. Stone, S.J. Shaw, M. Song, Effect of carbon nanotubes on the curing dynamics and network formation of Cyanate Ester resin. J. Polym. Res. **20**, 106–117 (2013b). https://doi.org/10.1007/s10965-013-0106-8

C. Lu, L. Yuan, Q. Guan, G. Liang, A. Gu, Optimizing ply pattern and composition of layered composites based on cyanate Ester, carbon nanotube and boron nitride: toward ultralow dielectric loss and high energy storage. J. Phys. Chem. C **122**, 5238–5247 (2018). https://doi.org/10.1021/acs.jpcc.7b12117

W.E. Mahmoud, F.S. Al-Hazmi, G.H. Al-Harbi, Wall by wall controllable unzipping of MWCNTs via intercalation with oxalic acid to produce multilayers graphene oxide ribbon. Chem. Eng. J. **281**, 192–198 (2015). https://doi.org/10.1016/j.cej.2015.06.081

S. Rakesh, C.P. Sakthidharan, M. Sarojadevi, P.R. Sundararajan, Monomer self-assembly and organo-gelation as a route to fabricate Cyanate Ester resins and their nanocomposites with carbon nanotubes. Europ. Polym. J. **68**, 161–174 (2015). https://doi.org/10.1016/j.eurpolymj.2015.04.023

T. Ramanathan, A.A. Abdala, S. Stankovich, D.A. Dikin et al., Functionalized graphene sheets for polymer nanocomposites. Nat. Nanotechnol. **3**, 327–331 (2008). https://doi.org/10.1038/nnano.2008.96

S. Stankovich, D.A. Dikin, G.H.B. Dommett et al., Graphene based composite materials. Nature **442**, 282–286 (2006). https://doi.org/10.1038/nature04969

W. Sun, W. Sun, M.R. Kessler et al., Multifunctional properties of Cyanate Ester composites with SiO_2 coated Fe_3O_4 fillers. ACS Appl. Mater. Interface **5**, 1636–1642 (2013). https://doi.org/10.1021/am302520e

Y. Wang, Z.X. Shi, J. Yin, Unzipped multiwalled carbon nanotubes for mechanical reinforcement of polymer composites. J. Phys. Chem. C **114**, 19621–19628 (2010). https://doi.org/10.1021/jp107151e

X. Wang, J. Jin, M. Song, Cyanate Ester resin/graphene nanocomposite: Curing dynamics and network formation. Europ. Polym. J. **48**, 1034–1041 (2012). https://doi.org/10.1016/j.eurpolymj.2012.03.012

N. Yang, X. Qi, D. Yang, M. Chen, Y. Wang, L. Huang, O. Grygoryeva, P. Strizhak, A. Fainleib, J. Tang, Improved mechanical, anti-UV irradiation, and imparted luminescence properties of cyanate ester resin/unzipped multiwalled carbon nanotubes/europium nanocomposites. Materials **14**, 4244 (2021). https://doi.org/10.3390/ma14154244

Z. Zhang, Z. Sun, J. Yao, D.V. Kosynkin, J.M. Tour, Transforming carbon nanotube devices into nanoribbon devices. J. Am. Chem. Soc. **131**, 13460–13463 (2009). https://doi.org/10.1021/ja9045923

Part II
Recent Advances in Studying Phthalonitrile Nanocomposites

Chapter 6
Phthalonitrile Composites with POSS Nanoparticles

Abstract The results of the physical study of a series of phthalonitrile nanocomposites with 0.5 wt.% of functionalized POSS nanoparticles are of four types. Their nanostructure, dynamics, relaxation, elastic and thermal properties were examined by TEM, mid-IR, far-IR, EDXS, DMA, DSC, and TGA techniques. Polymerization at temperatures not exceeding 300 °C resulted in the completion of the process by ~70–80% only, and post-curing at higher temperatures increased the degree of polymerization to 90–95%. Far-IR spectra confirmed the main contribution of phthalocyanine cycles to the matrix structure. TEM images and histograms of Si content in the nanovolumes of the composites (EDXS data) indicated a quasi-regular, at the first approximation, distribution of POSS nanoparticles in the amorphous matrix. The introduction of POSS nanoparticles and post-curing led to some suppression of the matrix dynamics and increasing T_g from 380–410 °C to 446 °C for neat matrix and to 520–560 °C for the nanocomposites, with the manifestation of dynamic heterogeneity in the glass transition. A complete suppression of the glass transition as well as the constancy of the dynamic modulus $E' \approx 3.2$ GPa over the range from 20° to 560 °C was observed after heating nanocomposites in an inert atmosphere. The total thermal stability of the nanocomposite up to ~ 400 °C and the "switching on" of thermo-oxidation processes in an air medium from 550 °C were registered by TGA.

6.1 Synthesis and Spectroscopic Control of Molecular Structure and Mobility

Recently, Bershtein, Fainleib et al. (Bershtein et al. 2019a, b, c, d; Yakushev et al. 2103) synthesized and performed a comprehensive physical study of a series of phthalonitrile (PhN) nanocomposites based on polymer matrices derived from bisphenol A-based phthalonitrile (BAPhN) and silicon-containing nanoparticles of different nature and with differently modified surfaces. The latter were reactive (relative to nitrile groups) amino- or epoxy-functionalized 3-D POSS nanoparticles and 2-D montmorillonite (MMT) nanolayers (see also Chap. 7). The nanostructure,

V. A. Bershtein and P. N. Yakushev, *High-Temperature Polymer Nanocomposites Based on Heterocyclic Networks from Nitrile Monomers*, Springer Series in Materials Science 334, https://doi.org/10.1007/978-3-031-32943-2_6

molecular dynamics, and properties of these nanocomposites were examined by TEM, mid-IR, far-IR, and EDXS, as well as using DMA, DSC, and TGA techniques.

The used monomer was 2,2-bis[4-(3,4-dicyanophenoxy)phenyl] propane (bisphenol A phthalonitrile, BAPhN). Figure 6.1 shows the names and, schematically, chemical formulas of the functionalized POSS nanoparticles used in Bershtein et al. 2019a. The BAPhN/nanoparticle mixtures (with 0.5 wt.% nanoparticles) were cured in a programmable oven by stepwise heating (8 h at 260 °C, 8 h at 280 °C, and 10 h at 300 °C). The films of 0.5 mm thickness were obtained. The matrix crosslinking, i.e., the degree of completion of the polymerization process, was controlled by a decrease in the intensity of the absorption band featuring the stretching vibrations of –C≡N groups with a maximum of ~2230 cm^{-1}. It was found that the above curing schedule resulted in an incomplete polymerization of BAPhN (Fig. 6.2). In view of this, the authors (Bershtein et al. 2019a) performed a number of post-curing procedures at different temperatures with estimating the attained polymerization degrees (PD) of –C≡N groups. Table 6.1 shows the IR spectroscopic results of matrix PD evaluation after heat treatment (post-curing) of nanocomposites with both POSS and MMT nanoparticles. Post-curing was produced either by treatment for different times at 340° or 370 °C or by heating up to 430° or 530 °C with the rate of 3 °C min^{-1} directly in the DMA setup.

The results of curing and post-curing experiments with spectroscopic PD evaluations showed the following. The curing at temperatures not exceeding 300 °C resulted only in a partial, far from the completion, polymerization reaction. Thus, in the case of a neat BAPhN matrix PD = 83%, and in the case of POSS-containing nanocomposites, PD was even lower: PD = 71.5–74.5%. This result was quite understandable: it was obviously due to spatial obstacles for the completion of the polymerization process and limited mobility in the densely cross-linked matrix network; mobility was additionally suppressed in the presence of covalently embedded nanoparticles (see below). At the same time, post-curing, which took place at higher temperatures and increased mobility of the network parts, allowed us to increase additionally the degree of cross-linking up to the limiting values of PD of around 90–95%. Below are the results obtained in experiments performed on both cured and post-cured samples.

In several studies, the authors investigated spectroscopically the molecular structure of phthalonitrile matrices (Li et al. 2017; Sheng et al. 2014, 2013; Kumar et al. 1993; Zou et al. 2012; Wu et al. 2017; Ji et al. 2016). It turned out that these polymer networks have a complex structure that varies depending on the composition, curing conditions, and other factors. Generally, polymerization of phthalonitrile monomer can occur with the formation of triazine, phthalocyanine, and isoindoline heterocycles. Thus, only triazine (Li et al. 2017; Sheng et al. 2014) or isoindoline (Kumar et al. 1993), or phthalocyanine heterocycles (Zou et al. 2012) were found in the phthalonitrile networks, although the other authors indicated the formation of both triazine and phthalocyanine (Wu et al. 2017) or isoindoline and phthalocyanine (Ji et al. 2016), or isoindoline and triazine heterocycles (Sheng et al. 2013).

Figure 6.3 shows schematically the course of reactions in the formation of the BAPhN/APIB-POSS nanocomposite matrix, or more precisely, the transformation of the initial monomer into three types of heterocycles which form the matrix network:

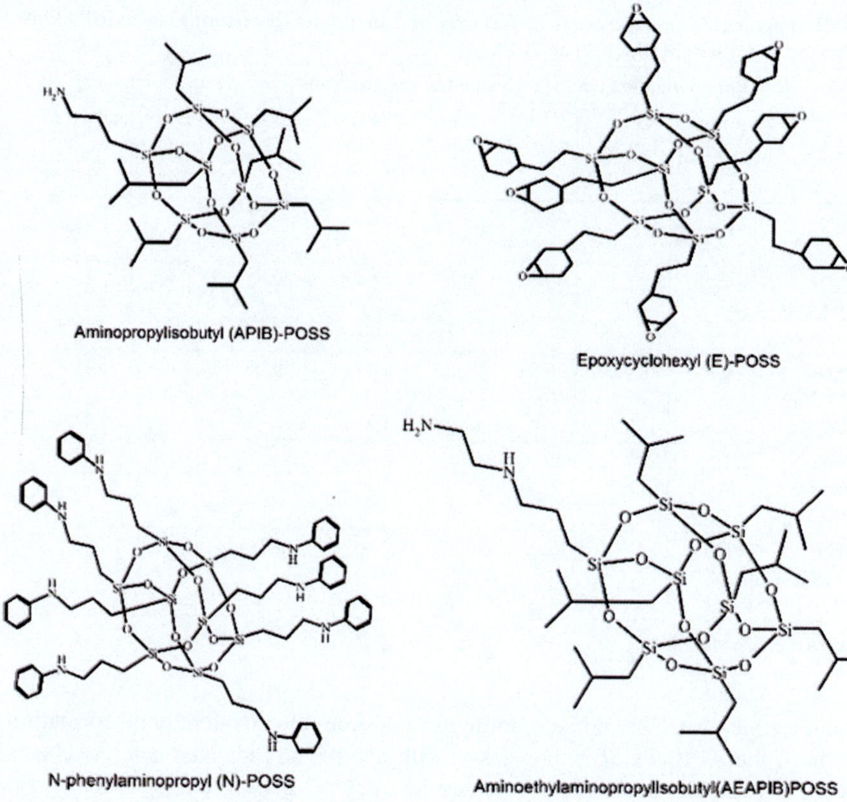

Fig. 6.1 Formulas of organically modified 3-D POSS nanoparticles with amino- or epoxy-functional groups (obtained from Hybrid Plastics, USA) which were used for preparing the phthalonitrile nanocomposites (Bershtein et al. 2019a)

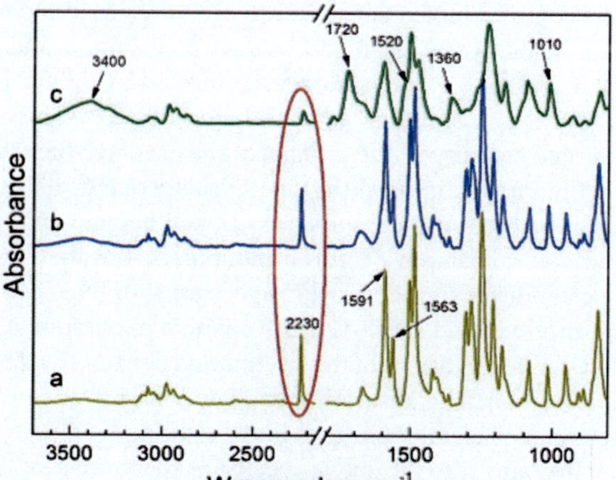

Fig. 6.2 IR spectra of BAPhN monomer (**a**), intermediate, partially polymerized product (**b**), and cured polymer matrix (**c**). The absorption band at 2230 см^{-1} corresponds to –C≡N groups (Bershtein et al. 2019a)

Table 6.1 Polymerization degrees attained at curing and for different post-curing modes in BAPhN-based nanocomposites (Bershtein et al. 2019a)

Sample composition (wt. %)	Polymerization degree (%) after the specified post-curing procedures							
	Initial state[a]	0.5 h, 340 °C	3.5 h, 340 °C	8 h, 340 °C	1 h, 370 °C	5 h, 370 °C	DMA heating 3°/min, up to	
							430 °C	530 °C
BAPhN matrix	83.2	84.2	90.4	91.4	93	93.1	–	–
BAPhN/ MMT I.31PS (99.5/0.5)	87.7	88.2	90	92.2	94.3	94.1	93.9	93.2
BAPhN/ MMTI.30E (99.5/0.5)	89	88	90.9	94.2	94.8	95.5	–	95.3
BAPhN/ APIB-POSS (99.5/0.5)	71.5	71	89	90	–	–	–	–
BAPhN/ E-POSS (99.5/0.5)	74.5	74	82.5	84.5	89	88.7	–	–

[a]Samples cured up to 300 °C

phthalocyanine, triazine, and isoindoline. In addition, the covalent bond formation reaction of the APIB-POSS nanoparticle with one of the indicated heterocycles is shown. Thus, the formation of the indicated heterocycles in the matrix was confirmed spectroscopically. Table 6.2 shows the values of characteristic IR absorption bands of polymerized phthalonitrile network; the spectra of the phthalonitrile nanocomposite in the mid-IR region are considered also in Sect. 9.7.

Thus, the polymerization of the BAPhN monomer generates a polymer network of a complex molecular structure. However, as found in our experiments, phthalocyanine heterocycles still made the main contribution and play the leading role (Bershtein et al. 2019a, b). The latter was confirmed by the far-IR spectra given below, as well as by the determining role of phthalocyanine cycles in the formation of the ultra-high thermal stability of these nanocomposites (see Sect. 9.7).

Figure 6.4 compares the far-IR spectra of BAPhN-based and CER-based materials. Both matrices are densely cross-linked heterocyclic polymer networks obtained by the polymerization of nitrile monomers, but their molecular structures and far-IR spectra differ significantly. In spectrum 4 of the CER-based matrix network, a broad absorption band at 50–120 cm^{-1} with a maximum at 84 cm^{-1}, corresponding to the low-angle vibrations of the triazine and benzene cycles (Bershtein and Ryzhov 1994), is observed. The absorption band at 238 cm^{-1}, which is also present, is ascribed to torsional skeletal vibrations in the matrix.

In contrast to the above spectrum, spectrum 1 of the polymerized BAPhN matrix contains only a low-intensity absorption band determined by the mentioned cycles;

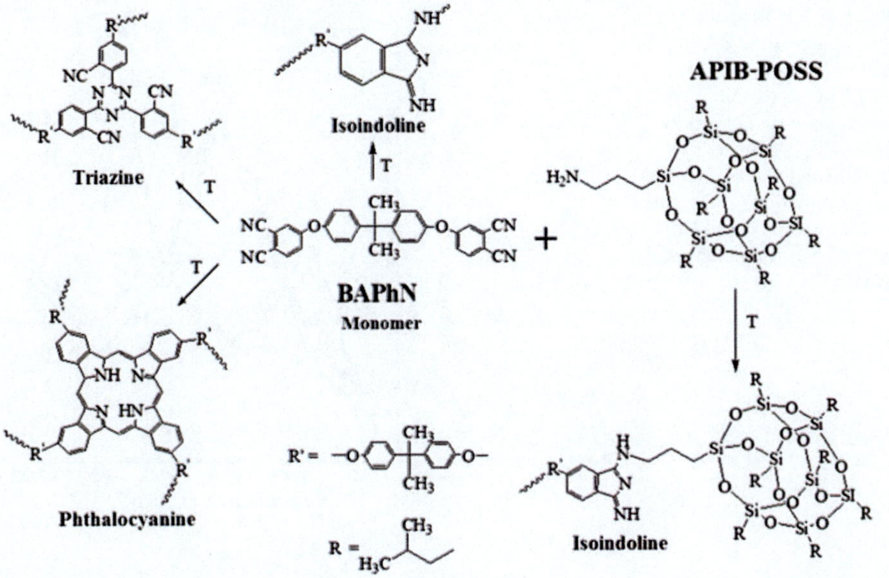

Fig. 6.3 Heterocycles formed at BAPhN polymerization within the matrix at the synthesis of BAPhN/APIB-POSS nanocomposite, and the scheme of covalent bonding between the APIB-POSS nanoparticle and the matrix (Bershtein et al. 2019a)

Table 6.2 Some characteristic absorption bands in the mid-IR spectra of phthalonitrile materials (Bershtein et al. 2019a)

Absorption band (cm^{-1})	Band assignment
2230	–C≡N stretching vibrations
~1360	Triazine cycles vibrations
~1520	Ditto
1010	Metal-free phthalocyanine cycles vibrations
2900–3400	NH stretching vibrations in the phthalocyanine
~1720	Isoindoline cycle vibrations

this can be caused by both a sharp decrease in the content of triazine heterocycles and the suppression of their mobility. The suppressed mobility is also confirmed by a large decrease in the absorption intensity in the spectral region of skeletal vibrations (200–300 cm^{-1}). The main feature of spectrum 1 of the BAPhN-based matrix is an intense absorption band at 155 cm^{-1} with small contributions from the 131, 182, and 224 cm^{-1} bands. Based on the experiments of Kobayashi (Kobayashi 1970), who obtained the far-IR spectra of phthalocyanine and its derivatives, it was concluded that the decisive contribution to the molecular structure of the BAPhN-based matrix is made by phthalocyanine heterocycles (Bershtein et al. 2019a).

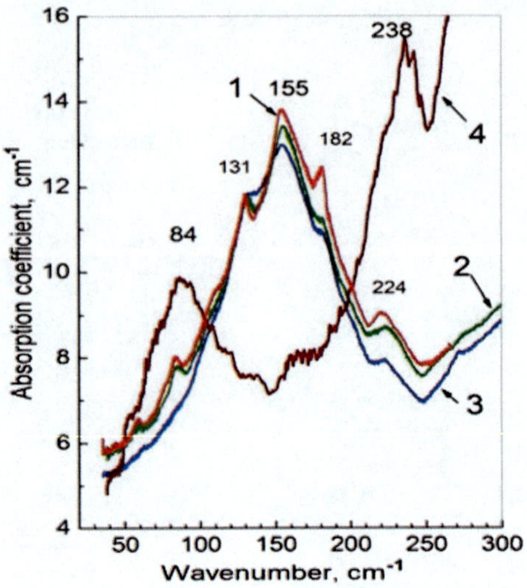

Fig. 6.4 Far-IR spectra of the cured pristine BAPhN matrix (1), BAPhN/APIB-POSS (99.5/0.5 wt.%) nanocomposite before (2) and after post-curing by heating up to 430 °C (3), and (for comparison) of the cured pristine DCBE based CER network (4) (Bershtein et al. 2019a)

In addition, Fig. 6.4 shows that the introduction of POSS nanoparticles into the matrix and especially the post-curing led to some additional suppression of the matrix dynamics and, possibly, the disappearance of triazine cycles: this is evidenced by both the reduced intensities of the absorption bands and the absence of an obvious manifestation of absorption bands in the region of 50–100 cm^{-1} in the spectrum.

6.2 Nanostructure

Figure 6.5 shows the typical TEM image and energy-dispersive X-ray (EDX) spectrum obtained for the nanocomposite with 0.5 wt. % of organically functionalized POSS nanoparticles. One can see that within the noise level, this nanocomposite looks as nanohomogeneous, i.e., there are no nanoclusters in it. This suggests good dispersion of nanoparticles at the molecular level in the amorphous matrix. It is essential that each EDX spectrum, obtained usually from a 10 nm diameter area, indicated the presence of Si atoms in each nanovolume of the composite. Such efficient dispersion of functionalized POSS nanoparticles could be explained by a rapid reaction of "embedding" of POSS particles (molecules) into the forming matrix network; this prevented the aggregation of nanoparticles.

It should be emphasized that in TEM measurements, individual POSS particles are difficult to distinguish; only POSS clusters of 2–5 nm in size and larger are distinctly fixed. In view of this, the authors (Bershtein et al. 2019a) used a statistical approach, obtaining large series of EDX spectra. Taken from a set (about 100) of randomly chosen spots, and then constructing the histograms of Si content in nanovolumes of the composite. In each case, the electron beam was focused on a spot about 2 nm

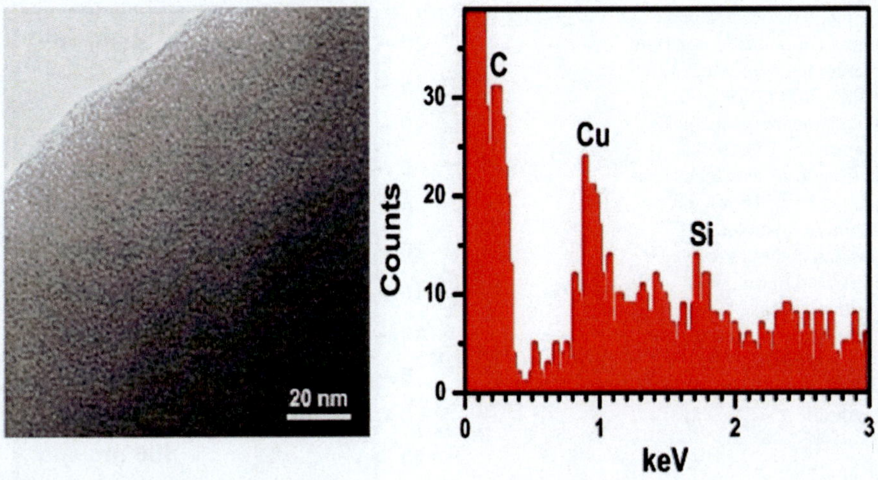

Fig. 6.5 A typical TEM image and EDX spectrum characteristic of the BAPhN/APIB-POSS (99.5/0.5 wt.%) nanocomposite. A signal from the copper supporting grid is also present in the spectrum (Bershtein et al. 2019a)

or 10 nm, or 100 nm in diameter to thus obtain information from nanovolumes of different scales.

Figure 6.6 presents histograms of Si nanodistribution in E-POSS-containing nanocomposites using these three experimental designs. In addition to the histograms, Gaussian contours are also shown here. Their widths are predetermined by the stochastic nature of the characteristic X-ray signal generated by the interaction of the electron beam with the material. This is actually the width of the instrument function characterizing the standard error. Therefore, Gaussian contours correspond to imaginary contours of the quasi-regular distribution of POSS nanoparticles in the nanocomposite matrix.

As can be seen from Fig. 6.6, in the case of "wide" electron beams (100 nm or 10 nm) the histograms of Si distribution in the nanovolumes of the composite coincide quite closely with the Gaussian contours, and their maxima coincide. This coincidence leads to a significant conclusion: the Si atoms, i.e., POSS units, are distributed quasi-regularly, at the first approximation, in the amorphous BAPhN matrix. Meanwhile, narrowing the electron beam down to 2 nm results in a significant broadening of the histogram and its substantial deviation from the Gaussian contour.

Figure 6.7 illustrates schematically the results presented. Indeed, the estimates showed that at 0.5 wt.% of the modified POSS nanoparticles in the nanocomposite, the average distance between the nanoparticles is approximately 7–8 nm. Naturally, therefore, there is no significant deviation of the histogram shape from the Gaussian contour when the electron beam width is 100 nm or 10 nm. At the same time, it is understandable why the measurements at 2-nm beam width resulted in their

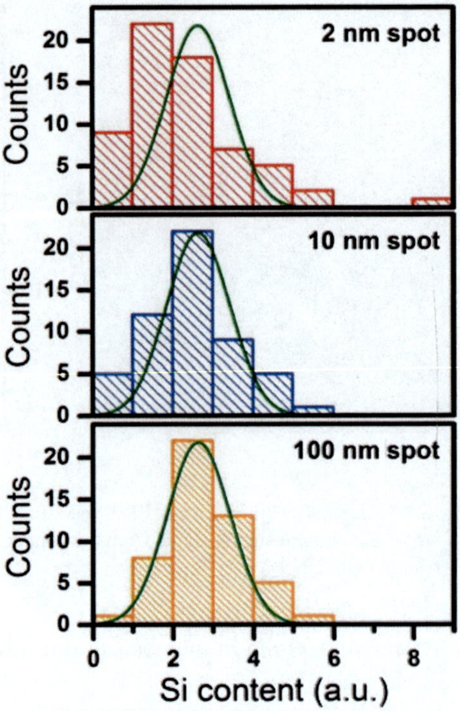

Fig. 6.6 Histograms of Si content (in arbitrary units) in different nanovolumes of the BAPhN/E-POSS(99.5/0.5 wt.%) nanocomposite as estimated by EDXS. The electron beam was focused at spots of 2 nm, 10 nm, or 100 nm in diameter. Gaussian contours correspond to the hypothetical case of a quasi-regular Si distribution in the amorphous BAPhN-based matrix (Bershtein et al. 2019a)

obvious divergence, with the manifestation of both the "POSS-depleted" and "POSS-enriched" nanoregions (Fig. 6.7). However, this latter result does not contradict the conclusion about the quasi-regular distribution of POSS units in the matrix.

6.3 Glass Transition and Dynamic Mechanical/Thermal Behavior

As shown above in Fig. 6.4, when 0.5% APIB-POSS is incorporated into the BAPhN matrix and post-curing is performed, spectral changes indicate some suppression of vibrational dynamics in the matrix network. One could expect that this should lead to corresponding changes in the relaxation and elastic properties of the nanocomposites. Indeed, this was confirmed in the experiments performed by the DMA method in the tensile mode at 1 Hz (Bershtein et al. 2019a). The loss modulus E'', the dynamic (storage) modulus E', and the mechanical loss factor $\tan\delta = E''/E'$ as functions of temperature were measured. The glass transition temperatures, T_g's were determined at $\tan\delta$ peak maxima. The experiments were performed at a heating rate of 3 °C min^{-1}, in an air atmosphere or under flowing nitrogen conditions. Usually, two heating cycles were performed for each sample: the first cycle, from 20° to 430 °C,

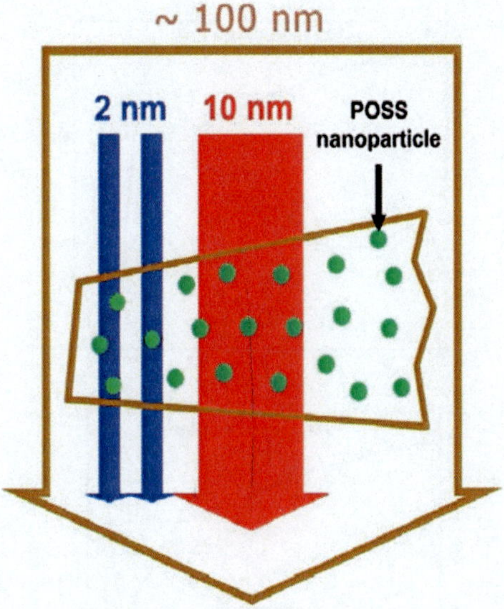

Fig. 6.7 A scheme of TEM experiments with the transmittance of an electron beam with different widths, 2 nm, 10 nm, or 100 nm, through the POSS-containing nanocomposite sample (Bershtein et al. 2019a)

characterized the mechanical properties of the cured composites (this was simultaneously a post-curing procedure), and the second heating cycle, from 20° to 570 °C, characterized the properties of the post-cured samples. In some experiments, a third heating cycle was also performed, from 20° to 570 °C, in a nitrogen flow environment.

Some DMA information obtained for BAPhN/POSS(0.5wt.%) nanocomposites is presented in Figs. 6.8 and 6.9, and the T_g values and modulus E' values at several temperatures, estimated by DMA for both cured and post-cured samples, are presented in Table 6.3. As shown in Table 6.3, in the case of initially cured, partially crosslinked matrices, the effect of POSS nanoparticles on matrix dynamics is relatively small: the BAPhN matrix and two POSS-containing nanocomposites have a glass transition temperature $T_g = 380–390$ °C, while the other two nanocomposites have $T_g = 410$ °C.

At the same time, post-curing led to a more densely cross-linked matrix and, as a result, to a significant improvement in the thermal characteristics of the nanocomposites. Additionally, a complication of the glass transition relaxation peak contour was observed and dynamic heterogeneity occurred. The glass transition peak contour varied depending on the type of functionalized POSS nanoparticles used. Thus, for a neat post-cured matrix, $T_g = 446$ °C. From the data presented in Table 6.3 and Fig. 6.8, we can see the following. In all four types of nanocomposites studied, the post-curing also leads to a significant increasing T_g. However, in two cases only $T_g \approx 460$ °C is registered, while the other two nanocomposites also show multiple T_g peaks at ~500°, 520°, and 560 °C. Obviously, all four peaks correspond to different in magnitude "constrained dynamics" effects in the nanovolumes of the composites. Thus, the post-curing of the BAPhN matrix led to an increase in its T_g by 57°, and

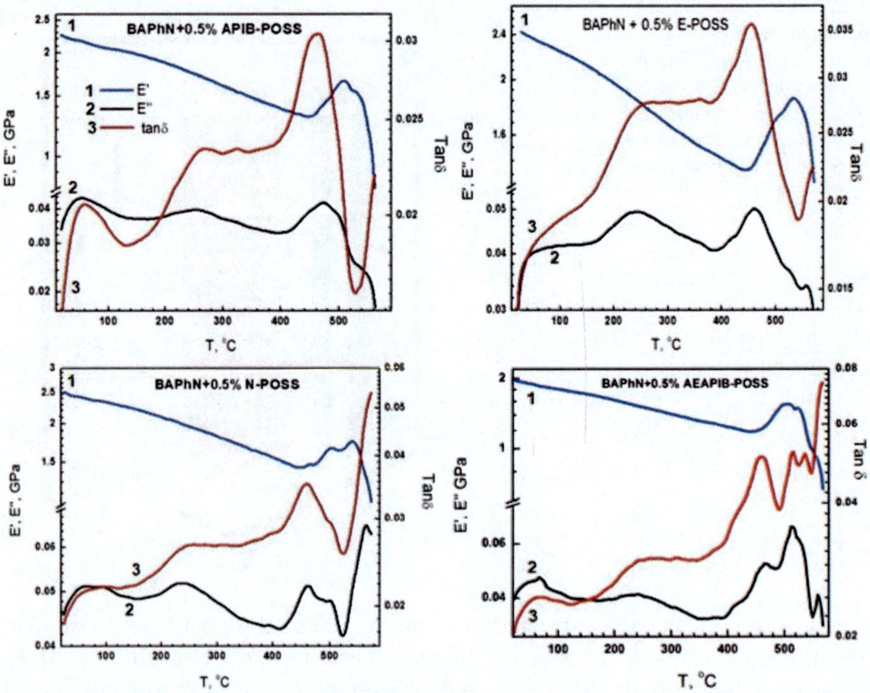

Fig. 6.8 DMA data, 1 Hz: dynamic modulus E'(T), loss modulus E'' (T) and tanδ (T) dependencies obtained for the post-cured samples of four BAPhN-POSS nanocomposites with 0.5 wt.% of differently functionalized POSS nanoparticles (Bershtein et al. 2019a)

the introduction of POSS nanoparticles led, additionally, to its increase from 446° up to 560 °C.

In the study (Bershtein et al. 2019a), the activation barriers of relaxation transitions in BAPhN-POSS nanocomposites were estimated by DMA method. For this purpose, tanδ (T) plots were measured at frequencies of 0.1, 1, and 10 Hz. The activation energies of sub-T_g relaxations, e.g., at 50–100° and 200–250 °C, were shown to be typical for non-cooperative, Arrhenius processes, e.g., for the latter transition the activation energy $Ea = 166$ kJ mol^{-1}.

At the same time, these experiments revealed an anomalous behavior of the glass transition. It turned out that, within the accuracy of its estimate (3°), T_g value remained unchanged at different frequencies; this corresponded to the value of the "apparent" activation energy $E_a > 1000$ kJ/mol. This allowed us to conclude that due to the highly cooperative nature of the transition (in this case of a densely cross-linked heterocyclic network), the glass transition behaves as a quasi-phase transition in a solid.

As can be seen from Table 6.3 and Fig. 6.8, the value of the dynamic modulus E' of the BAPhN-POSS(99.5/0.5 wt.%) nanocomposites varies significantly not only with temperature, but also due to post-curing and the incorporation of one or another type

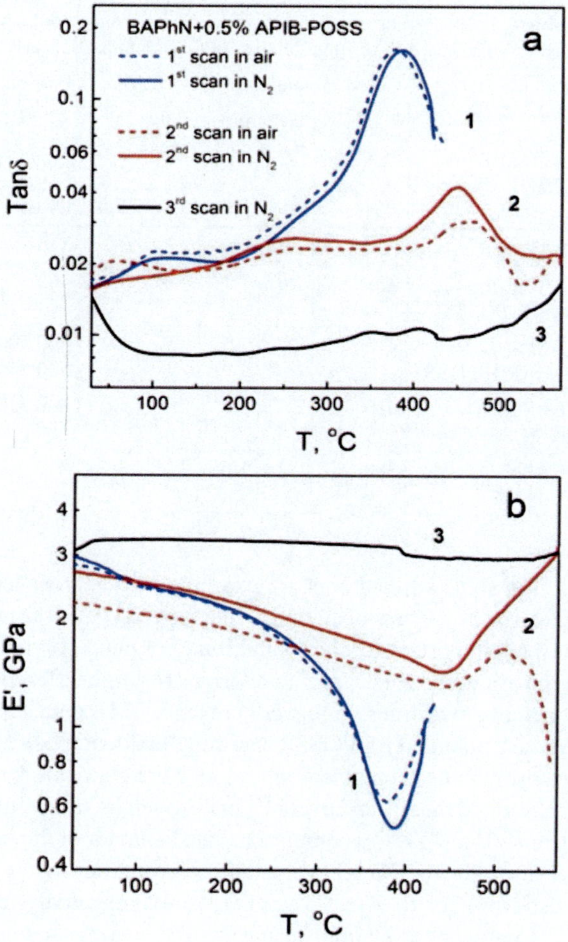

Fig. 6.9 DMA data (1 Hz), air versus nitrogen medium: tanδ (T) (**a**) and dynamic modulus E'(T) (**b**) curves for the BAPh/APIB-POSS(99.5/0.5 wt.%) nanocomposite (Bershtein et al. 2019a)

of functionalized POSS nanoparticles. At 30 °C the modulus E′ = 2.0–2.8 GPa for cured nanocomposites; the introduction of nanoparticles increases it to a small extent, although post-curing may even lead to some reduction of the modulus value. This was not so at high temperatures. In these cases, the post-cured composites always showed, in particular, an increase in modulus with increasing temperature from its maximum value during post-curing (430 °C) up to about 500 °C (Fig. 6.8) due to attaining a higher matrix crosslink density. At 400 °C, the modulus of some post-cured nanocomposites was 1.5–3.0 times higher than that of the cured samples and 1.5 times higher than that of the neat matrix. At 500 °C, in some cases, the modulus was higher than at 400 °C, e.g., 1.7 GPa for the BAPhN/N-POSS nanocomposite (Table 6.3). Decreasing E′ at T >500 °C was associated in part with the onset of thermo-oxidative degradation of the materials.

Table 6.3 Glass transition temperatures and dynamic modulus values for BAPhN-based nanocomposites with 0.5 wt.% of functionalized POSS nanoparticles (Bershtein et al. 2019a)

Sample	Cured samples					Post-cured samples					
	T_g (°C)	Dynamic modulus E′ (GPa) at 1 Hz and T(°C)				T_g (°C)	Dynamic modulus E′ (GPa) at 1 Hz and T(°C)				
	DMA	30	200	300	400	DMA	30	200	300	400	500
BAPhN matrix	389	2.4	1.6	1.3	0.5	446	1.5	1.3	1.1	0.9	1.2
BAPhN/ APIB-POSS	378	2.8	2.1	1.4	0.7	462	2.2	1.9	1.6	1.4	1.6
BAPhN/ AEAPIB-POSS	410	2.7	2.5	2.0	1.0	460 500 520 560	1.9	1.3	1.1	0.9	1.3
BAPhN/ N-POSS	410	2.7	2.5	2.0	1.0	460 520 560	2.5	2.1	1.8	1.6	1.7
BAPhN/ E-POSS	388	2.1	1.6	1.2	0.5	459	2.4	2.0	1.7	1.5	1.7

For some applications, for example in the aerospace field, comparative studies of material properties in air and in an inert, oxygen-free atmosphere are of interest. This is most important for the evaluation of organic polymeric materials operating at high temperatures, which are characterized by large difference in their thermal and thermo-oxidative stabilities. Figure 6.9 presents DMA data, temperature dependences of tanδ and E', obtained for POSS containing nanocomposite in air and the flowing nitrogen environments. Three heating cycles were conducted, of which cycles I and II in two mentioned media, and cycle III in nitrogen medium only. The first cycle, with heating from 20° to 430 °C, characterized the behavior of the initially cured composite. Cycle II from 20° to 570 °C evaluated the properties of the post-cured sample. Cycle III, also from 20° to 570 °C, was performed repeatedly in a nitrogen environment.

Figure 6.9 shows that during the first heating (scanning) to 430 °C, the influence of the oxidizing medium on the relaxation spectrum and on the temperature dependence of the dynamic modulus is negligibly small: in both mediums $T_g \approx 380$ °C, and the modulus E' decreases sharply from 300 °C, slightly increasing then at 400–430 °C due to post-curing. The influence of the oxidizing air environment becomes noticeable in scan II, when the experimental curves obtained in the air medium and in an inert atmosphere diverge substantially. The greatest effect in this case is E' behavior at temperatures over 500 °C, with the beginning of the intensive thermo-oxidative degradation processes (see below): E' decreases from 1.7 GPa to 0.8 GPa in the air but increases up to 3.2 GPa in the nitrogen atmosphere.

Finally, a particularly interesting and nontrivial result was obtained when scan III was performed up to 560 °C in an inert nitrogen atmosphere. Figure 6.9 shows that in this case there is actually complete suppression ("disappearance") of the relaxation spectrum, including glass transition, as well as the constancy of the dynamic modulus $E' \approx 3.2$ GPa over the entire temperature range studied, from 20° to 560 °C. To our knowledge, the achievement of a constant value of the elastic modulus over such

Fig. 6.10 TGA: mass versus temperature plots obtained in air or nitrogen medium for the BAPhN/E-POSS (99.5/0.5 wt.%) nanocomposite (Bershtein et al. 2019a)

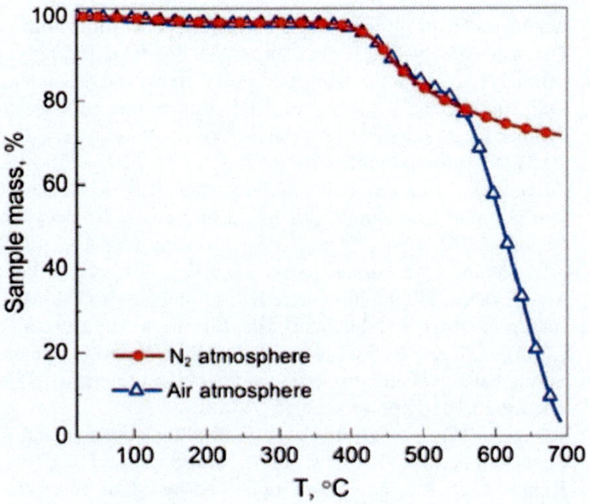

a wide temperature range has been observed for polymeric materials in our work (Bershtein et al. 2019a) for the first time.

Figure 6.10 presents the results of thermogravimetric analysis (TGA) of the BAPhN-POSS nanocomposite films when heated between 20° and 700 °C at a rate of 20 °C min^{-1} in air and nitrogen mediums. One can see the absolute thermal stability of the nanocomposite up to ~ 400 °C; the mass loss of 1.5–2 wt. % under these conditions is due to the removal of absorbed water. Weak thermal degradation started from about 430 °C. Up to ca. 550 °C, when about 20% mass removal occurred, the curves of thermal and thermo-oxidative degradation coincided (no air influence), and the samples in both cases retained their integrity. At the same time, it is the temperature 550 °C that turned out to be a specific "temperature bifurcation point": the TGA plots diverged sharply from this temperature due to the "switching on" of intensive thermo-oxidation processes during the tests in the air environment.

As a result, further destructive processes when heated to 700 °C in an inert atmosphere were small, while in an air medium there was actually complete decomposition of the material. As shown below (see Sect. 9.7), the superhigh thermal stability of phthalonitrile nanocomposites is explained by the high content of exceptionally thermostable phthalocyanine heterocycles in the matrix network.

References

V.A. Bershtein, V.A. Ryzhov, Far-infrared spectroscopy of polymers. Adv. Polym. Sci. **114**, 43–121 (1994). https://doi.org/10.1007/BFb0008694

V.A. Bershtein, A.M. Fainleib, P.N. Yakushev, D.A. Kirilenko, K.G. Gusakova, D.A. Markina, O.G. Melnychuk, V.A. Ryzhov, High-temperature phthalonitrile nanocomposites with silicon-based nanoparticles of different nature and surface modification: structure, dynamics and properties. Polymer **165**, 39–54 (2019a). https://doi.org/10.1016/j.polymer.2019.01.020

V.A. Bershtein, A.M. Fainleib, P.N. Yakushev, D.A. Kirilenko, K.G. Gusakova, D.A. Markina, O.G. Melnychuk, V.A. Ryzhov, High-temperature phthalonitrile/amino-MMT nanocomposites: synthesis, structure and properties. Express Polym. Lett. **13**, 656–672 (2019b). https://doi.org/10.3144/expresspolymlett.2019.55

V.A. Bershtein, A.M. Fainleib, P.N. Yakushev, D.A. Kirilenko, O.G. Melnychuk, Super-heat resistant polymer nanocomposites based on hetercyclic networks: structure and properties. Phys. Solid State **61**, 1494–1501 (2019c). https://doi.org/10.1134/S1063783419080080

V.A. Bershtein, A.M. Fainleib, P.N. Yakushev, D.A. Kirilenko, K.G. Gusakova, O.G. Melnichuk, V.A. Ryzhov, High temperature hybrid phthalonitrile-MMT nanocomposites: synthesis, structure, properties. In: Abstracts of the European polymer congress, NAN-C38, Crete, p. 192

S. Ji, P. Yuan, J. Hu, R. Sun, K. Zeng, G. Yang, A novel curing agent for phthalonitrile monomers: curing behaviors and properties of the polymer network. Polymer **84**, 365–370 (2016). https://doi.org/10.1016/j.polymer.2016.01.006

T. Kobayashi, The far infrared spectra of phthalocyanine and its metal derivatives. Spectrochim. Acta, Part A **26**, 1313–1322 (1970). https://doi.org/10.1016/0584-8539(70)80037-X

D. Kumar, U.D. Razdan, A.D. Gupta, heat-resistant polymers from melt-processable bisimido-bisphthalonitriles. J. Polym. Sci., Part A: Polym. Chem. 31, 797–804 (1993). https://doi.org/10.1002/pola.1993.080310326

X. Li, B. Yu, D. Zhang, J. Lei, Z. Nan, Cure behavior and thermomechanical properties of phthalonitrile–polyhedral oligomeric silsesquioxane copolymers. Polymers **9**(8), 334 (2017). https://doi.org/10.3390/polym9080334

H. Sheng, X. Peng, H. Guo, X. Yu, C. Tang, X. Qu, Q. Zhang, Synthesis and thermal properties of a novel high-temperature alkyl-center-trisphenolic-based phthalonitrile polymer. Mater. Chem. Phys. **142**, 740–747 (2013). https://doi.org/10.1016/j.matchemphys.2013.08.033

H. Sheng, X. Peng, H. Guo, X. Yu, K. Naito, X. Qu, Q. Zhang, Synthesis of high-performance bisphthalonitrile resins cured with self-catalyzed 4-aminophenoxy phthalonitrile. Thermochim. Acta **577**, 17–24 (2014). https://doi.org/10.1016/j.tca.2013.12.010

Z. Wu, J. Han, N. Li, Z. Weng, J. Wang, X. Jian, Improving the curing process and thermal stability of phthalonitrile resin via novel mixed curing agents. Polym. Int. **66**(6), 876–881 (2017). https://doi.org/10.1002/pi.5328

P.N. Yakushev, V.A. Bershtein, A.M. Fainleib, D.A. Kirilenko, O.G. Melnychuk, Ultra-heat resistant nanocomposites based on heterocyclic networks: structure, properties, the origin of thermal stability. J. Phys.: Conf. Ser. 2103 012113 (2021). https://doi.org/10.1088/1742-6596/2103/1/012113

Y. Zou, J. Yang, Y. Zhan, X. Yang, J. Zhong, R. Zhao, X. Liu, Effect of curing behaviours on the properties of poly(arylene ether nitrile) end-capped with phthalonitrile. J. Appl. Polym. Sci. **125**(5), 3829–3835 (2012). https://doi.org/10.1002/app.36691

Chapter 7
Phthalonitrile/Montmorillonite Nanocomposites

Check for updates

Abstract The results of physical studies of molecular structure, nanostructure, relaxation, elastic and thermal properties of a group of hybrid phthalonitrile nanocomposites with different contents, from 0.03 to 5 wt.%, of functionalized MMT nanolayers, using TEM, EDXS, DMA, DSC, TGA, and mid-IR/far-IR spectroscopy are shown. Far-IR spectra confirmed the main contribution of phthalocyanine heterocycles to the molecular structure of nanocomposites and the effect of constraining matrix dynamics after "embedding" MMT. The degree of exfoliation of MMT blocks in nanocomposites increased with decreasing their content: at 0.03–0.1 wt.% MMT single nanolayers were formed. For cured materials, neat matrix and nanocomposites, T_g = 380–390 °C but after post-curing T_g of the matrix increased to 446 °C, whereas the nanocomposites exhibited dynamic heterogeneity in the glass transition and their T_g's reached 540° and 570 °C. After treatment at high temperatures in an inert medium, the complete suppression of the glass transition and constant modulus E' = 3 GPa in the range from 20° to ~ 600 °C were observed for the nanocomposites. The nanocomposites were stable when heated up to 400 °C, regardless of the nature of the environment. The incorporation of MMT nanolayers led to a substantial improvement in the resistance to thermo-oxidative degradation in the air environment. The char residue at ~ 700 °C increased from 7% for the neat matrix to 45% for the nanocomposite with 2% MMT.

7.1 Structure

Among the inorganic nanoparticles, 2-D silicate montmorillonite (MMT) nanolayers with a thickness of about 1 nm, already mentioned above, occupy a special position. This is due to the good mechanical properties of MMT nanolayers, their high aspect ratio, and the accessibility and low cost of this mineral. Such nanolayers are obtained from blocks of natural minerals by their exfoliation in the medium of a polymerizing matrix. Recently, Bershtein, Fainleib et al. (Bershtein et al. 2019a, 2019b, 2019c, 2019d; Yakushev et al. 2103) have synthesized and performed a comprehensive physical study of the molecular structure, nanostructure, and relaxation, elastic and

V. A. Bershtein and P. N. Yakushev, *High-Temperature Polymer Nanocomposites Based on Heterocyclic Networks from Nitrile Monomers*, Springer Series in Materials Science 334, https://doi.org/10.1007/978-3-031-32943-2_7

thermal properties of a group of phthalonitrile nanocomposites with different MMT contents, from 0.03 to 5 wt.%.

As in the case of POSS-containing phthalonitrile nanocomposites (see Chap. 6), a monomer of 2,2-bis[4-(3,4-dicyanophenoxy)phenyl] propane (bisphthalonitrile of bisphenol A, BAPhN monomer) was used to form a heterocyclic polymer matrix (see above Fig. 6.3). One of three types of organically modified montmorillonite (MMT) with the trade names Nanomer® I.30E.

Nanomer® I.31PS (amino-MMT), or Nanomer® I.28E (Nanocor Inc., USA, see Fig. 7.1) was introduced. The content of MMT in the nanocomposites was varied from 0.03 wt.% to 5 wt.% in the case of Nanomer® I.30E in works (Bershtein et al. 2019b, 2019c, 2019d; Yakushev et al. 2103); in the other two cases, 0.5 wt.% of nanoparticles only were introduced (Bershtein et al. 2019a). In the case of Nanomer® I.30E, the nanoparticle involved 70–75 wt.% MMT and 25–30 wt.% octadecylamine (Fig. 7.1).

The structure, matrix dynamics, thermal, relaxation, and elastic properties of the BAPhN/MMT nanocomposites were characterized for the first time using simultaneously TEM, EDXS, DMA, DSC, TGA methods, and FTIR and Far-IR spectroscopy. The experiments were performed both in an air atmosphere and in a flowing nitrogen medium. As shown above in Table 6.1, step curing of the BAPhN/MMT nanocomposites at 260–300 °C resulted in PD = 87–89%, and post-curing at 370–530 °C provided some additional increase in matrix cross-linking density to PD = 95%. IR spectra of these nanocomposites confirmed the same complex molecular structure of their heterocyclic matrix as in the case of POSS-containing nanocomposites (see Chap. 6). The functionalized MMT nanoparticles were incorporated into the matrix network due to the high reactivity of nitrile –C≡N groups relative to the amine groups of nanoparticles.

Far-IR spectroscopic measurements (Bershtein et al. 2019a) showed that in the case of POSS-containing phthalonitrile nanocomposites based on the BAPhN matrix,

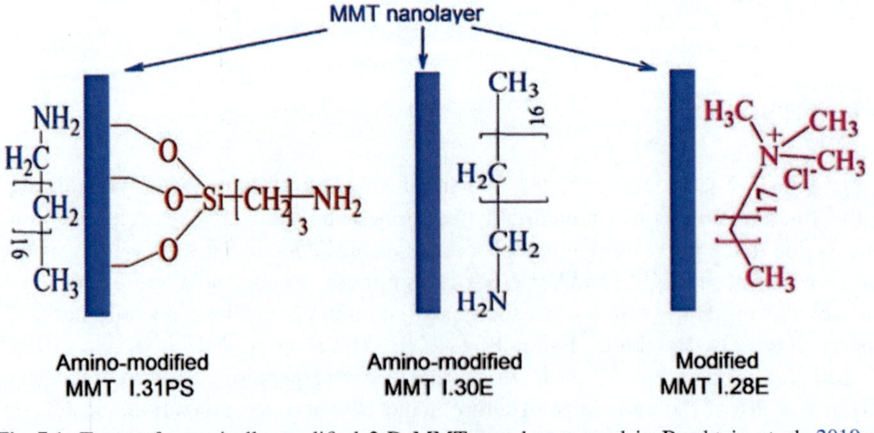

Fig. 7.1 Types of organically modified 2-D MMT nanolayers used in Bershtein et al. 2019a; Bershtein et al. 2019b; Bershtein et al. 2019c; Bershtein et al. 2019d; Yakushev et al. 2103)

the phthalocyanine heterocycles mainly contributed to the molecular structure of the latter (Fig. 6.4). A similar result was obtained in the case of MMT-containing nanocomposites. In addition, in a similar way, the introduction of reactive amino-MMT nanolayers into this matrix and the post-curing procedure led to noticeable changes in the spectrum due to the "constraining dynamics" effect (Bershtein et al. 2019b).

Figure 7.2 presents the far-IR spectra in the region from 20 cm^{-1} to 300 cm^{-1} for the cured neat BAPhN-based matrix and for the BAPhN/MMT1.30E (2 wt.%) nanocomposite in the cured and post-cured states. Really, similar to that which was observed in the spectra of POSS-containing nanocomposites (Fig. 6.4), the most intense absorption band, at 155 cm^{-1}, as well as low-intensity contributions of bands with maxima at ~130 cm^{-1}, 180 cm^{-1}, and 220 cm^{-1} in the spectrum of neat BAPhN-based matrix are observed. This confirms again the major contribution of phthalocyanine heterocycles to the matrix network. The spectrum of the matrix contains also low-intense absorption bands of the triazine heterocycles at 54 cm^{-1} and 84 cm^{-1}.

The introduction of 2 wt.% nanoparticles and the post-curing procedure resulted in three small but discernable changes in the far-IR spectrum (Fig. 7.2): (a) a decrease in absorption intensity; (b) a shift of the peak maximums by a few cm^{-1} toward higher wave-numbers; and (c) a practical disappearance of the absorption peaks at wave numbers less than 100 cm^{-1}. The latter effect may be related both to the

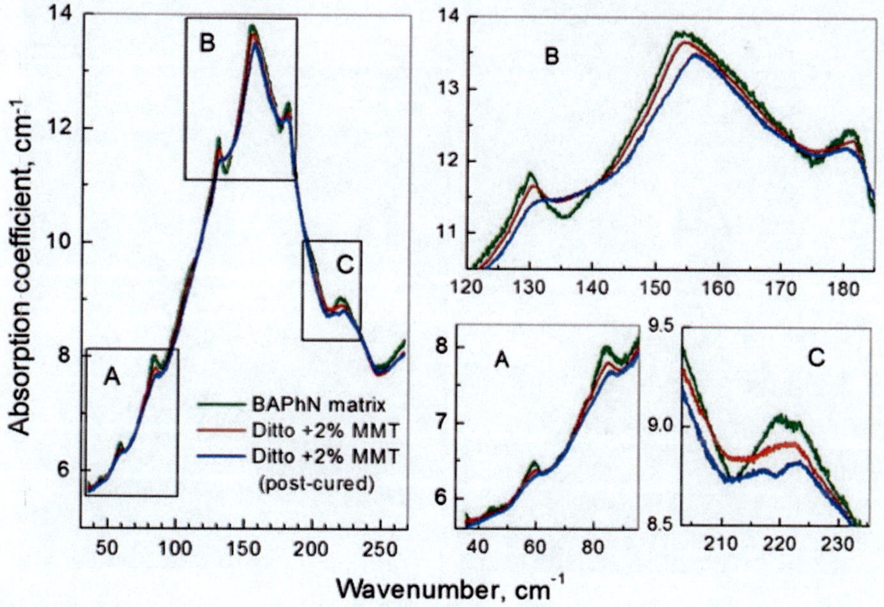

Fig. 7.2 Far-IR spectra of the cured neat BAPhN-based matrix and the BAPhN/MMT1.30E (2 wt.%) nanocomposite in the cured and post-cured states (enlarged images are shown on the right) (Bershtein et al. 2019b)

transformation of the triazine cycles and to their strongly suppressed mobility in the matrix network. On the whole, these spectral changes correspond to the effect of constraining network dynamics.

TEM analysis combined with EDXS spectral measurements allowed us to characterize the exfoliation degree of the MMT blocks (stacks) and the distribution of the amino-MMT nanolayers in the matrix. EDX spectra were obtained by focusing electron beams on nanocomposite sites with a diameter of 2 nm. This allowed direct verification of observed structural features, and control of the local elemental composition (Si, Mg, Al) in the nanovolumes of the composite, and to confirm spectroscopically the presence of single 1-nm thickness MMT nanolayers.

Figure 7.3 shows the typical TEM micrographs taken from 2-nm spots for the BAPhN/MMT1.30E nanocomposites with different contents of nanofillers. One can see that the degree of exfoliation of the MMT blocks in the matrix is highly dependent on the MMT content and can be very different.

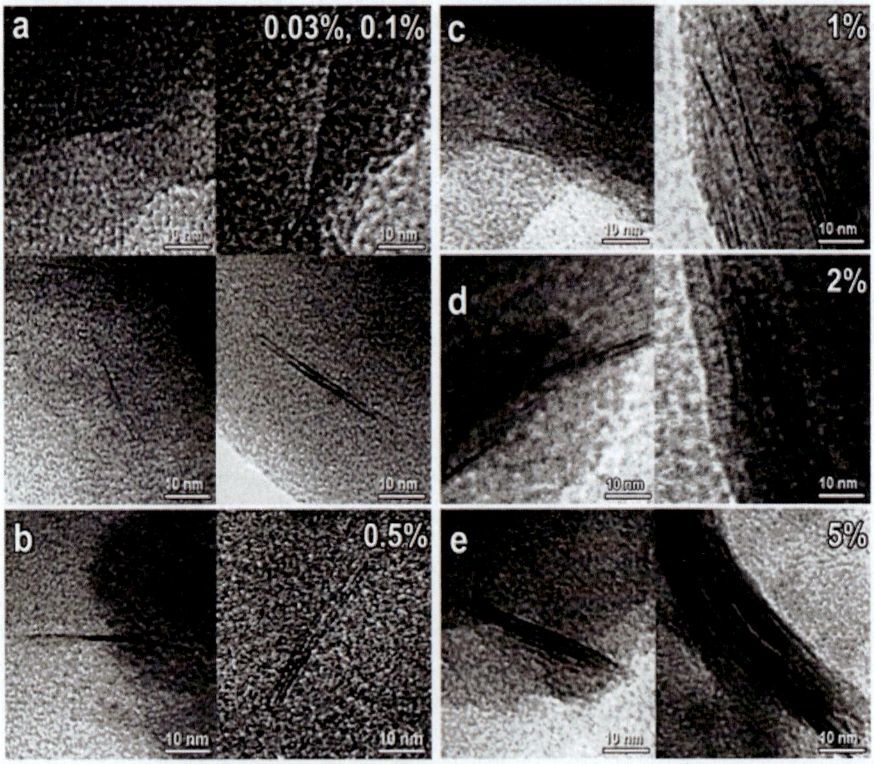

Fig. 7.3 TEM micrographs obtained for the BAPhN/MMT 1.30E nanocomposites with different indicated contents of nanofillers. For each composition typical images are shown on the left, and rarely registered ones are shown on the right, at the content of **a** 0.03% and 0.1%, **b** 0.5%, **c** 1%, **d** 2%, and **e** 5 wt.% MMT (Bershtein et al. 2019b)

The highest degree of exfoliation of MMT blocks is attained at their ultra-low content in the nanocomposite: at 0.03–0.1 wt.%. In this case, the matrix contains predominantly single MMT nanolayers and only rarely two-layer MMT formations. In the nanostructure of the composite with 0.5 wt.% MMT, single MMT nanolayers are also present, but along with this, thin, three- to four-layer MMT stacks are also detected. An interesting pattern of MMT exfoliation can be seen in the TEM micrograph of the nanocomposite with 1 wt.% MMT. In this case, a 100 nm photo clearly shows completely different degrees of exfoliation: single nanolayers, two-layer formations, and five- to ten-layer stacks of MMT. Finally, at 2–5 wt.% MMT, a relatively low level of MMT stack exfoliation was observed: TEM recorded almost exclusively only thin and "thick" MMT stacks of about 5 nm and 20 nm thickness, respectively (Fig. 7.3).

Figure 7.4 gives examples of typical EDX spectra obtained using an electron beam focused on a 2 nm diameter area from a matrix, a single MMT nanolayer, or a stack of MMT nanolayers.

Some peculiarities of the manifestation of these spectra should be noted. First, the most intense signals from Si and other MMT elements in the case of a single

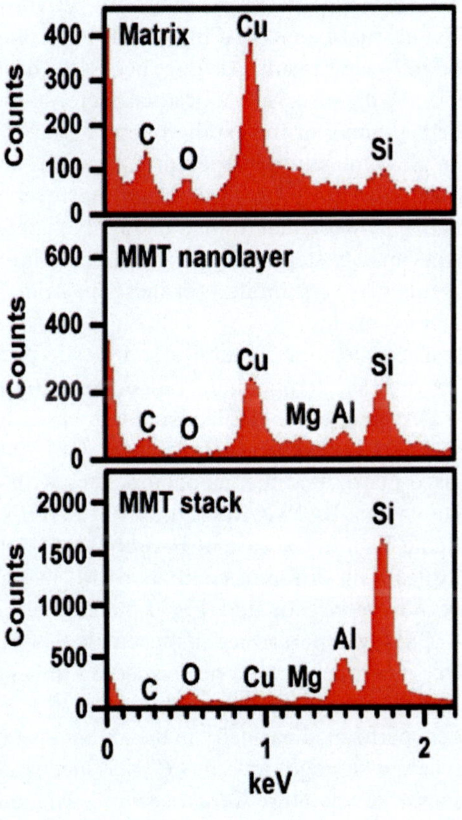

Fig. 7.4 EDX spectra of the BAPhN/MMT1.30E nanocomposites taken from 2 nm spots in diameter at focusing electron beam at polymer matrix, at single MMT nanolayer, or at MMT stack. A signal from Cu is not related to the sample (Bershtein et al. 2019b)

MMT nanolayer take place if the nanolayer is oriented in parallel to the electron beam. Since the site on which the electron beam was focused is still larger than the thickness of the MMT nanolayer, the spectrum will inevitably also contain signals from matrix elements (C, O). Nevertheless, EDX spectra of single MMT nanolayers, including signals of basic elements of MMT (Si, Al, and Mg), allow us to confirm that the dark lines observed in the TEM images really belong to MMT nanolayers of about 1 nm thickness. Secondly, the stacks of MMT layers give more intense signals of Si, Al, and Mg, although the small signals from the matrix are present in the spectrum in this case as well. Thirdly, in the EDX spectrum of the matrix, there are minor signals from the MMT elements because the electron beam can cross some layers of the MMT, located at an obtuse angle relative to it and not visible in the TEM image. At last, the Cu signal in the EDX spectrum refers not to the sample but to the supporting grid.

7.2 Dynamic Mechanical Analysis

DSC experiments (Bershtein et al. 2019b) were performed for samples of neat BAPhN matrix and BAPhN/MMT nanocomposite with 0.5 wt.% MMT and showed the following result. At four successive heating cycles up to 370°, 400°, 430°, and 450 °C, the effects of a gradual increase in T_g due to the post-curing process and the beginning of the exothermic material degradation were observed. For example, the glass transition temperature of cured samples $T_g = 386$ °C increased to 416 °C after post-curing. In these cases, however, the high degree of crosslinking in the matrix network determined its highly suppressed mobility and therefore very small heat capacity steps at T_g. In general, the information extracted from the DSC experiments was very limited for these materials. DMA was much more informative for their evaluation.

The relaxation dynamics and elastic properties of BAPhN/MMT nanocomposites over the temperature range from 20° to 570 °C were evaluated using DMA in Bershtein et al. 2019a; Bershtein et al. 2019b; Bershtein et al. 2019c; Bershtein et al. 2019d; Yakushev et al. 2103). DMA has been performed for the neat BAPhN-based matrix and the nanocomposites with 0.5 wt.% of differently modified MMT nanolayers: BAPhN/MMT I.31 PS, BAPhN/MMT I.30E, and BAPhN/MMT I.28E (Table 7.1). In the case of BAPhN/MMT I.30E compositions, a series of nanocomposites with different contents of MMT nanolayers ranging from 0.03 wt.% to 5 wt.% were investigated (Fig. 7.5 and Table 7.1).

DMA was performed in the tensile mode at 1 Hz, at the heating rate of 3 °C·min^{-1}. The experiments were performed on film samples with a thickness of 0.5 mm; the working part of the test samples was 20 × 5 × 0.5 mm^3 in size. The measurements were performed basically in the air but also under N_2 atmosphere. Loss modulus E'', dynamic (storage) modulus E', and mechanical loss factor $tan\delta = E''/E'$ as the functions of temperature were measured. The glass transition temperatures, T_g's, at *tanδ (T)* peaks maxima (or, sometimes, also at $E''(T)$ peaks maxima) were determined.

Table 7.1 Glass transition temperatures and dynamic modulus values for BAPhN-based nanocomposites with 0.5 wt.% MMT nanolayers (Bershtein et al. 2019a)

Sample	Cured samples					Post-cured samples					
	DMA Tg (°C)	Dynamic modulus *E*' (GPa) at 1 Hz)				DMA Tg (°C)	Dynamic modulus E' (GPa) at 1 Hz				
		30	200	300	400 °C		30	200	300	400	500 °C
BAPhN pristine matrix	389	2.4	1.6	1.3	0.5	446	1.5	1.3	1.1	0.9	1.2
BAPhN/MMT I.31PS	387	2.8	2.0	1.3	0.5	457 515	1.9	1.5	1.3	1.2	1.3
BAPhN/MMT I.30E	387	3.2	2.3	1.5	0.5	455 500 540	1.5	1.4	1.2	1.1	1.7
BAPhN/MMT I.28E	379	2.5	1.6	1.0	0.5	454 520	1.5	1.3	1.1	0.9	1.1

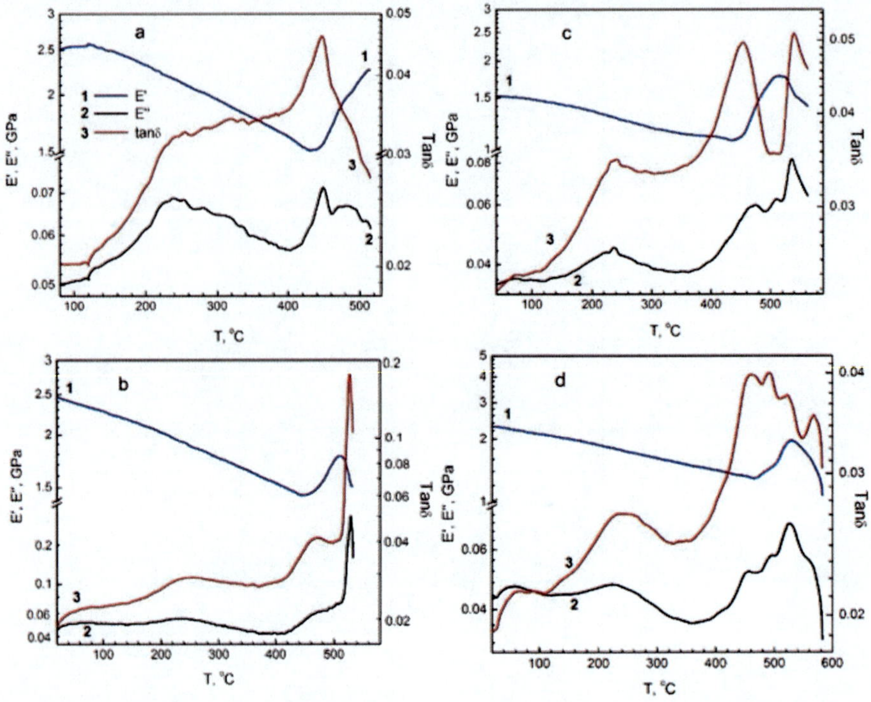

Fig. 7.5 DMA data (1 Hz) obtained for post-cured samples of the neat BAPhN matrix (**a**) and BAPhN-based nanocomposites with 0.03 wt.% (**b**), 0.5 wt.% (**c**), and 1.0 wt.% MMT 1.30E (**d**) (Bershtein et al. 2019b, 2019c)

Three heating cycles were carried out: the first cycle from 20° to 430 °C characterized the properties of the composites cured at step heating from 20° to 300 °C; the second one, from 20° to 570 °C, estimated the post-cured state of the sample, and then sometimes the third heating cycle from 20° to 570 °C was carried out in flowing nitrogen medium only, in the absence of oxidizing processes.

One can see from Table 7.1 that for the cured neat BAPhN matrix and three types of cured nanocomposites based thereon, T_g, as estimated by DMA, is equal to 380–390 °C (The relaxation spectra of all cured samples are not presented here; one example only is shown below in Fig. 7.6). It means that the influence of nanoparticles on the matrix dynamics was practically absent in these cases of partly crosslinked matrix networks. Relatively small changes in the E' values are also observed in this case.

However, after the post-curing procedure by heating up to 430 °C, i.e., during the second scanning the relaxation spectra changed dramatically (Tables 7.1 and 7.2, Fig. 7.5).

In general, these nanocomposites with the most densely crosslinked heterocyclic matrix (PD ≈ 95%, see Table 6.1) are characterized by a decrease in the level of mechanical losses in the studied temperature range, especially in the glass transition,

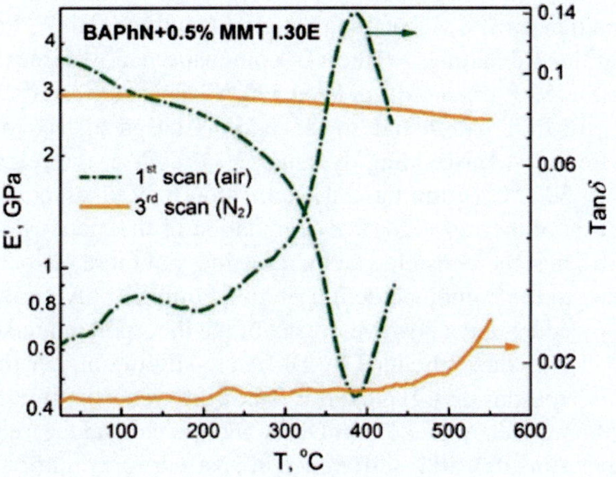

Fig. 7.6 DMA data, 1 Hz: *tanδ* (T) and dynamic modulus *E'* (T) plots obtained for the cured sample of BAPhN/ MMT1.30E (0.5 wt.%) nanocomposite during the first scanning in air medium and during rescanning in nitrogen environment after heating in this inert medium up to 570 °C (Bershtein et al. 2019c)

Table 7.2 Glass transition temperatures of post-cured BAPhN/MMT1.30E nanocomposites (DMA, 1 Hz) (Bershtein et al. 2019c)

MMT1.30E, wt.%	T_g, °C
0	446
0.03	460, 529
0.1	460, 500, 540
0.5	455, 540
1	460, 493, 520, 570
2	460, 539
5	470

and by a significant improvement in thermal characteristics. Thus, in the glass transition peaks *tanδ* changed from ≈ (0.1–0.2) for the cured samples down to 0.04–0.05 for the post-cured ones. The relaxation spectra distinctly demonstrate three broad overlapping relaxation regions, namely sub-T_g relaxations at ~50–100° and 200–250 °C, and a glass transition range extending in general for the post-cured ones from about 400° to 570 °C. Moreover, the glass transition of the studied post-cured nanocomposites acquires a complicated character. The relaxation spectra of these nanocomposites differ from the spectrum of the pristine BAPhN matrix, and their contours depend on the type of functionalization (organic modification) of the MMT surface and MMT content.

A single glass transition peak with $T_g = 446$ °C for the neat post-cured BAPhN matrix but the dissimilar, complicated spectral contours for the post-cured nanocomposites were observed. One can see a different dynamic heterogeneity in the composites' glass transition including its doublet structure, or three-four overlapping glass transition peaks (see also Table 7.1). Dynamic heterogeneity in the glass transition was most pronounced in the nanocomposite with 1% MMT (Fig. 7.5 d). Therefore, DMA data indicate the presence of matrix nanovolumes with relatively slight influence of MMT on the post-cured matrix (with $T_g \approx 460$ °C) and different "constrained dynamics" effects in composite nanovolumes; this resulted in $T_g \approx 470°$, 500°, 520°, 530–540° or even 570 °C. Thus, the satisfactory exfoliation of the modified MMT nanolayers in the BAPhN-based matrix may result in the pronounced effect of "constraining dynamics". Of course, the relaxation peaks at temperatures T > 500 °C within the glass transition may be associated, to some extent, with the onset of thermo-oxidative degradation of the matrix network (see next section).

Thus, we registered an increase in T_g of these nanocomposites by ~80° to 180 °C due to the combined action of an additional polymerization during the post-curing procedure and a positive impact of the incorporated nanoparticles. It should be noted that T_g values obtained by DMA are slightly higher than those measured by DSC. Two reasons may explain this fact: a difference in time conditions of the experiments (the frequency of 1 Hz in DMA and the equivalent frequency of $\sim 10^{-2}$ Hz in DSC experiments), and a difference in post-curing conditions: by scanning with the rates of 3 °C/min and 20 °C/min in DMA and DSC experiments, respectively.

Analysis of the relaxation spectra of nanocomposites indicates their following peculiarities (Fig. 7.5, Tables 6.3 and 7.1). The introduction of ultra-low amounts of MMT (0.03 wt.%) into the matrix, i.e., mainly of individual MMT nanolayers, led to their small suppressive influence on the dynamics in most of the matrix nanovolumes (shift of the relaxation peak from 446° to 460 °C) and simultaneously to the appearance of a second narrow peak with a maximum at 529 °C, characterizing probably T_g in the matrix nanovolumes which are directly localized near the MMT nanolayers. After introducing 0.1% MMT, the relaxation picture became more complicated, and the spectra with strongly overlapped peaks at ca. 460°, 500°, and 540 °C within the glass transition range may be seen. A double structure of the glass transition peak with the maxima at 455° and 540 °C is observed for the nanocomposite with 0.5% MMT. The most interesting glass transition manifestation was registered for the nanocomposite with 1% MMT: it is the complicated relaxation picture consisting of four overlapping peaks with close intensities and the maxima at 460°, 493°, 520°, and 570 °C. Of importance, this pronounced dynamic heterogeneity turned out to be in good accordance with the nanostructural data. Just, in this case, we observed simultaneous manifestation of different MMT exfoliation degrees including the single MMT layers, two-layers, and five-ten layers MMT stacks in the matrix within ~ 100 nm at the TEM image. At last, at 2% and especially at 5% MMT in the nanocomposites, the glass transition is simplified, and only doublet structure of the T_g relaxation peak is observed, at 460° and 539 °C in the former case, and only one peak with maximum at 470 °C at 5% MMT. It may be understood, again, proceeding from the structural viewpoint. The degree of dispersion (exfoliation) of MMT stacks sharply decreases in these cases. Therefore, only rather "thick" MMT stacks and nanovolumes of the matrix, relatively "free" from the influence of MMT nanolayers prevail in these nanocomposites. It means that the interfacial area in these nanocomposites and constrained dynamics effects may be much less than those at low MMT contents. These results are in the same line of interpretation as the anomalously large positive effects of ultra-low additions of POSS and SiO_2 units considered above.

Figure 7.5 and Table 7.1 also show peculiar changes in the dynamic (storage) modulus *E*' with temperature, different for cured and post-cured samples. For cured nanocomposites at 30 °C, the E' modulus increased from 2.4 GPa for neat matrix to 2.5GPa, 2.8GPa and (for BAPhN/MMT 1.30E) to 3.2 GPa. The value of *E*' decreased sharply, however, with increasing temperature to the same values of 0.5 GPa at 400 °C in all four cases. Then *E*' increased, to a small extent, while heating to 430 °C due to the post-curing process and attaining a higher matrix crosslink density (PD = 95%, see above).

Another data were obtained for the modules *E*' of post-cured nanocomposites, i.e., in a second temperature scanning (Table 7.1, Fig. 7.5). In this case, the introduction of MMT resulted in a certain decrease in *E*' values at 30 °C—from 2.4 GPa to 1.7 ± 0.2 GPa. There is no the unambiguous explanation of this effect so far. However, presumably it may be associated with a slight thermal degradation starting from ~430 °C at the first scanning (see the next section). At the same time, however, the post-cured nanocomposites showed a significant superiority in their high-temperature behavior. This consisted in a considerably slower decrease of E' with increasing temperature and a jump-like increase of E' in the temperature range of 430–500 °C. It is noteworthy that for the post-cured BAPhN/MMT I.30E nanocomposite E' at 500 °C was even higher than at 30 °C (Table 7.1). It was natural to observe decreasing E' at T > 500 °C due to thermo-oxidative degradation of the materials (see next section).

As for the BAPhN/POSS nanocomposites (see Fig. 6.9), in the case of the BAPhN/MMT nanocomposite a similar effect was also found, in general, unusual for polymeric materials. Figure 7.6 shows the DMA data, *tanδ*(T), and dynamic modulus *E*'(T) plots, obtained for the cured BAPhN/amino-MMT(0.5 wt.%) nanocomposite at the scanning in air medium up to 430 °C and rescanning in nitrogen medium after heating in this inert environment up to 570 °C. One can see again the complete suppression of the relaxation spectrum, the disappearance of the glass transition, and the practically constant value of modulus *E*' = 3 GPa in the temperature range from 20° to ~600 °C.

7.3 Thermal Stability

Thermogravimetric analysis (TGA) and derivative thermogravimetry (DTG) of neat BAPhN matrix and three BAPhN/MMT I.30E nanocomposites with 0.5, 2, and 5 wt.% nanofillers were performed under both air and inert (flowing nitrogen) atmospheres (Bershtein et al. 2019b, 2019c). Samples with masses ranging from 10 to 20 mg were heated from 20° to 690° or 890 °C with the rate of 20 $°C{\cdot}min^{-1}$. The results obtained are presented in Fig. 7.7 and Table 7.3. Their analysis allows us to make the following comments.

In general, the high thermal stability of these materials is quite obvious. Both the neat matrix and the nanocomposites are stable when heated up to 400 °C, regardless of the nature of the environment, and the observed mass loss of 1.5–2 wt. % is associated with the removal of absorbed water.

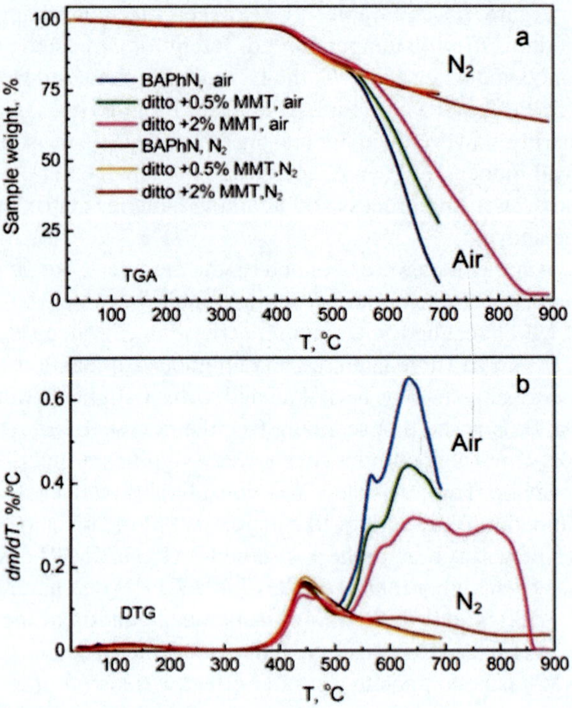

Fig. 7.7 TGA and DTG data obtained in air or nitrogen mediums for cured samples of neat BAPhN matrix and BAPhN/MMT I.30E nanocomposites with 0.5 or 2 wt.% nanofiller (Bershtein et al. 2019b, 2019c)

Table 7.3 Thermal stability and thermo-oxidative degradation of phthalonitrile materials (TGA/DTG results) (Bershtein et al. 2019b)

Air atmosphere										
Sample	T_d, °C			Temperature of maximal degradation rate, $T_{d\,max}$ (°C), at the stage				Char residue (wt.%) at T (°C)		
	5%	10%	20%	I	II	III	IV	690	750	890
BAPhN	426	460	539	446	567	636	–	7	–	–
BAPhN / MMT I.30E (0.5%)	429	462	549	445	569	636	–	28	–	–
BAPhN / MMT I.30E (2.0%)	437	479	566	444	589	645	780	45	28	~ 3
BAPhN / MMT I.30E (5.0%)	446	489	576	446	603	683	738	39	12	~ 5
Nitrogen atmosphere										
Neat BAPhN	432	463	557	445	–	–	–	73	–	–
BAPhN/MMT I.30E (0.5%)	431	461	559	448	–	–	–	74	–	–
BAPhN/MMT I.30E (2.0%)	430	463	577	444	618	–	–	72	69	64
BAPhN/MMT I.30E (5.0%)	445	487	608	449	497	613	–	76	74	71

Note The T_d and $T_{d\,max}$ are the temperatures corresponding to the indicated mass losses (5, 10, or 20 wt.%) and the maxima of DTG peaks, respectively

Slight thermal degradation begins at ~ 430 °C, when a mass loss of about 5% is registered, but this process remains virtually independent of the material composition and the type of ambient gas environment when heated up to 500 °C. At this temperature, the mass loss of the samples reaches about 20%, but the integrity of the film samples is maintained. At the same time, the temperature of 550 °C appears as the "temperature bifurcation point": starting from this temperature, the TGA plots obtained by measurements in air and nitrogen environments sharply diverge due to the "switching on" in the first case of intensive thermo-oxidative degradation processes. As a result, the TGA curve measured in air goes sharply downward, while further destructive processes when heated in an inert atmosphere are negligible up to 690 °C. Moreover, the data obtained showed that the studied nanocomposites retained their integrity even when heated in an inert atmosphere up to 890 °C (see Sect. 9.7). Thus, the satisfactory thermal stability of the nanocomposites under study, with retaining the material integrity, is observed at short-term heating up to 500 °C in air medium or up to ~ 900 °C in the oxygen-free medium.

Figure 7.7, a and Table 7.3 show that the thermal stability in a nitrogen atmosphere slightly depends or almost does not change after the introduction of MMT into the phthalonitrile matrix: the char residue at 690 °C is 72–76% in the case of a neat matrix as well as for the nanocomposites. At the same time, the incorporation of MMT nanolayers led to a substantial improvement in the resistance to thermo-oxidative degradation in the air environment. In particular, the char residue at ~ 700 °C increased from 7% for the pure matrix to 45% for the nanocomposite with 2% MMT.

It should be noted that the enhancement of thermal stability in polymer nanocomposites containing MMT has previously been observed in a number of studies (see, e.g., the review article (Leszczyńska et al. 2007)). The origin of this effect is, obviously, rather complicated: it presumably can be connected with the changes in the degradation kinetics due to the barrier effect, because of the limitations in oxygen diffusion, as a result of the suppression of the matrix dynamics which affects the reaction rate, etc.

As demonstrated above (see Chap. 2), TGA of CER/amino-MMT nanocomposites showed their intense thermal degradation even in an inert atmosphere already at 430–500 °C, when almost half of the sample mass was lost. In this case, the matrix contained only triazine and benzene rings. Korshak et al. (Korshak et al. 1974) showed that this degradation process corresponded to the breakage of bonds between the above rings in CER followed by an intensive decomposition of the cyanurate skeleton. Therefore, it was natural to assume that the superiority—the ultra-high thermal stability of phthalonitrile nanocomposites in an inert atmosphere—was determined by the leading contribution of ultra-stable phthalocyanine heterocycles to the molecular structure of the matrix. This assumption was confirmed by the spectroscopical data (see Sect. 9.7).

Table 7.3 and Fig. 7.7 b present the detailed experimental DTG data obtained at temperatures up to ~ 700–900 °C for neat matrix and BAPhN/MMT 130E nanocomposites with 0.5, 2, and 5 wt.% nanofiller. One can see that the DTG plot of the neat BAPhN-based matrix measured in a nitrogen atmosphere exhibits only one small degradation peak with a maximum at 445 °C. At the same time, in the nanocomposites with 2–5% MMT the destructive process continues to a small extent and at higher temperatures. Char residue can remain at ~ 70% up to a temperature of ~ 900 °C.

A much more complicated picture of the increased degradation of the BAPhN matrix and nanocomposites based thereon appears in the DTG plots obtained by measurements in the air. In this case, DTG curves include 3–4 strongly overlapping intense peaks; the maximum of the first of them is also fixed at 444–448 °C, and the last one at 780 °C. The introduction of MMT leads to a shift of the DTG curve toward higher temperatures and a decrease in the intensity of the degradation peaks.

Finally, incorporating MMT nanolayers resulted in some increase in the thermo-oxidative stability in air: in particular, the char residue at ~ 700 °C increased up to 45% compared with 7% for the neat matrix.

Thus, the above data and discussion clearly confirm that the satisfactory thermal stability of phthalonitrile nanocomposites, with preservation of material integrity, can be achieved up to temperatures of ~ 500 °C in air medium and up to 900 °C in oxygen-free atmosphere under short-term exposure to high temperatures. The reason for this superheat-resistant effect is also discussed below in Sect. 9.7. This allows us to certainly assume the leading role of this type of materials as the most thermostable No. 1 among polymer composites.

Two points should be noted. Of course, the above refers to the stability under relatively short-term thermal exposure to nanocomposites; prolonged exposure to ultra-high temperatures will lead to a certain decrease in the acceptable temperature of their exploitation. On the other hand, however, the experiments were carried out on film samples with a thickness of 0.5 mm. It is natural to assume, therefore, that in the case of massive parts made of nanocomposites, the negative effect of the oxidizing air environment will be lower than in the described experiments.

Thus, we conclude this chapter by emphasizing the following. Perhaps, it is phthalonitrile nanocomposites containing modified MMT nanolayers that are the most obvious candidate for the first place among thermostable nanocomposites. The role of uniquely thermostable phthalocyanine heterocycles in their structure is discussed in Sect. 9.7. This conclusion is supported by the following facts:

- unusually high thermal stability of the matrix itself,
- the strong effect of suppression of matrix dynamics by MMT nanolayers and the important role of interfacial dynamics,
- the manifestation of the pronounced dynamic heterogeneity in the glass transition,
- the achievement of the glass transition temperature of 570 °C, unique for polymeric matrices, and
- the display of short-term thermal stability, with retaining material integrity, when the material is heated up to 500 °C in air and up to 900 °C in an oxygen-free atmosphere.

References

V.A. Bershtein, A.M. Fainleib, P.N. Yakushev, D.A. Kirilenko, K.G. Gusakova, D.A. Markina, O.G. Melnychuk, V.A. Ryzhov, High-temperature phthalonitrile nanocomposites with silicon-based nanoparticles of different nature and surface modification: structure, dynamics and properties. Polymer **165**, 39 (2019a). https://doi.org/10.1016/j.polymer.2019.01.020

V.A. Bershtein, A.M. Fainleib, P.N. Yakushev, D.A. Kirilenko, K.G. Gusakova, D.A. Markina, O.G. Melnychuk, V.A. Ryzhov, High-temperature phthalonitrile/amino-MMT nanocomposites: synthesis, structure and properties. Express Polym Lett **13**, 656 (2019b). https://doi.org/10.3144/expresspolymlett.2019.55

V.A. Bershtein, A.M. Fainleib, P.N. Yakushev, D.A. Kirilenko, O.G. Melnychuk, Super-heat resistant polymer nanocomposites based on heterocyclic networks: structure and properties. Phys. Solid State **61**(8), 1494 (2019c). https://doi.org/10.1134/S1063783419080080

V.A. Bershtein, A.M. Fainleib, P.N. Yakushev, D.A. Kirilenko, K.G. Gusakova, O.G. Melnichuk, V.A. Ryzhov, High-temperature hybrid phthalonitrile-MMT nanocomposites: synthesis, structure, properties. In Abstracts of the European Polymer Congress, NAN-C38, Crete, p. 192

V.V. Korshak, P.N. Gribkova, A.V. Dmitrenko, A.G. Puchin, V.A. Pankratov, S. Vinogradova, Thermal and thermal-oxidative degradation of polycyanates. Polymer Science U.S.S.R., **16**, 15–23 (1974). https://doi.org/10.1016/0032-3950(74)90111-7

A. Leszczyńska, J. Njuguna, K. Pielichowski, J.R. Banerjee, Polymer/Montmorillonite nanocomposites with improved thermal properties: part II. Thermal stability of montmorillonite nanocomposites based on different polymeric matrices. Thermochimica Acta, **454**, 1 (2007). https://doi.org/10.1016/j.tca.2006.11.003

P.N. Yakushev, V.A. Bershtein, A.M. Fainleib, D.A. Kirilenko, O.G. Melnychuk, Ultra-heat resistant nanocomposites based on heterocyclic networks: structure, properties, the origin of thermal stability. J. Phys.: Conf. Series **2103**, 012113 (2021). https://doi.org/10.1088/1742-6596/2103/1/012113

Chapter 8
Phthalonitrile/Metal Oxide Nanocomposites

Abstract The structure and properties of three types of cured PhN/metal oxide nanocomposites with Al_2O_3, TiO_2, or ZnO nanoparticles of an average diameter of 60 nm, which were treated with aminopropyltrimethoxysilane GX-540 as coupling agent, are described. In all cases, a satisfactory dispersion of nanoparticles was achieved. A significant improvement in the thermomechanical properties of nanocomposites with respect to the properties of the neat matrix was shown. The experimental data obtained for the elastic modulus of the studied Al_2O_3-containing nanocomposites were compared with the Series, Halpin–Tsai, and Kerner theoretical models; the best match was shown for the Halpin–Tsai model. The barrier properties of the TiO_2-containing nanocomposites as anticorrosive protective coatings were shown. The UV–visible transmittance spectra of the ZnO-containing nanocomposites showed that adding ZnO nanoparticles improved the UV-shielding properties of the neat matrix.

8.1 Phthalonitrile/Alumina Nanocomposites

Derradji et al. (1549) and Shan et al. (2017) synthesized and studied two series of nanocomposites based on polymerized bisphenol-A phthalonitrile resin P(BAPh) and surface-modified Al_2O_3 nanoparticles. Al_2O_3 particles had an average diameter of about 60 nm. In (Derradji et al. 1549), they were treated with the aminopropyltrimethoxysilane (GX-540) coupling agent, and different amounts (3, 6, 9, 12, or 15 wt.%) of modified Al_2O_3 nanoparticles were introduced. In (Shan et al. 2017), the nanocomposites were based on phthalonitrile resin with 1, 3, or 5 wt.% of nitrile-functionalized Al_2O_3 nanoparticles.

To assure a satisfactory dispersion of the nanoparticles, the initial mixtures were vigorously stirred using a mechanical agitator for 4 h at the rate of 4000 rpm and then sonicated. These mixtures were cured by step heating at temperatures from 240 to 320 °C. FTIR, SEM, TEM, DMA, DSC, and TGA have been used for the characterization of these nanocomposites.

V. A. Bershtein and P. N. Yakushev, *High-Temperature Polymer Nanocomposites Based on Heterocyclic Networks from Nitrile Monomers*, Springer Series in Materials Science 334, https://doi.org/10.1007/978-3-031-32943-2_8

The FTIR spectra of untreated and silane-modified alumina nanoparticles indicated the successful grafting of the silane coupling agent onto Al_2O_3 nanoparticles (Derradji et al. 1549). The band in the range from 500 cm^{-1} to 900 cm^{-1} was characteristic of amorphous alumina, but a few bands were present only in the spectrum of the treated alumina nanoparticles. For example, the bands at 2944 cm^{-1} and 2843 cm^{-1} could be attributed to the asymmetric and symmetric $–CH_2$ stretching vibrations. The band at 1470 cm^{-1} was due to the $–CH_2$ bending vibrations, and the bands at 1590 cm^{-1} and 1073 cm^{-1} corresponded to the N–H and $Si–O–CH_3$ vibrations, respectively (Derradji et al. 1549).

TEM images of P(BAPh)/Al_2O_3 nanocomposites showed that the modified alumina nanoparticles were mostly well dispersed in the matrix even at their highest content (15 wt.%), although some agglomerates of nanoparticles smaller than 500 nm in size (also uniformly distributed in the matrix) were also present.

Figure 8.1 represents the DMA data on storage modulus and *Tanδ* relaxation peaks (along with the glass transition temperature T_g as estimated at the peaks maxima) obtained for a series of the cured BAPh/Al_2O_3 nanocomposites with different contents of the nanoparticles. It can be seen that at 50 °C, the storage modulus and T_g values of the neat matrix were equal to 1.5 GPa and 300 °C, respectively, and both these parameters increased significantly with the introduction of nanoparticles and the increase of their content in nanocomposites. So in the limit, at 15 wt,% silanized alumina nanoparticles, the modulus increased up to 3.4 GPa, and $T_g = 346$ °C. The significant superiority of nanocomposites over the neat matrix with respect to stiffness not only remains, but also increases at high temperatures, and at 300 °C the modulus of the neat matrix becomes negligibly small, while the modulus of the nanocomposite with 15 wt.% nanoparticles is close in its value to that of the neat matrix at 50 °C (Fig. 8.1a).

Figure 8.1b shows also that, as alumina nanoparticles were added, the relaxation peak of the *tanδ* was greatly reduced, which directly indicated that the nanoparticles covalently bound to the matrix were significantly suppressing the dynamics of the matrix. It was shown also (Derradji et al. 1549) the influence of the introduced alumina nanoparticles on the tensile strength, breaking strain, and microhardness of the nanocomposites under consideration. The tensile strength increased from 40 to 71 MPa after adding 15 wt.% silanized Al_2O_3 nanoparticles, although a very small negative effect of decreasing ductility was also observed. At last, with the introduction of Al_20_3 nanoparticles, the microhardness of the material increased from 400 to 500 MPa.

It should be noted that in this and a number of other works on phthalonitrile nanocomposites discussed in this chapter and Chap. 9, it was apparently possible to achieve fairly good nanoparticle dispersity and basically prevent the formation of their aggregates in the matrices. This made it possible to register the maximum positive effects when introducing rather high concentrations of nanoparticles (up to 6–15 wt.%). The effect of the special influence of ultra-low concentrations of nanoparticles, which was discussed in the previous chapters, was not tested in these works. At the same time, in the works discussed in this chapter and Chap. 9, the systems were cured only at temperatures not exceeding 320 °C and the post-curing procedure at the

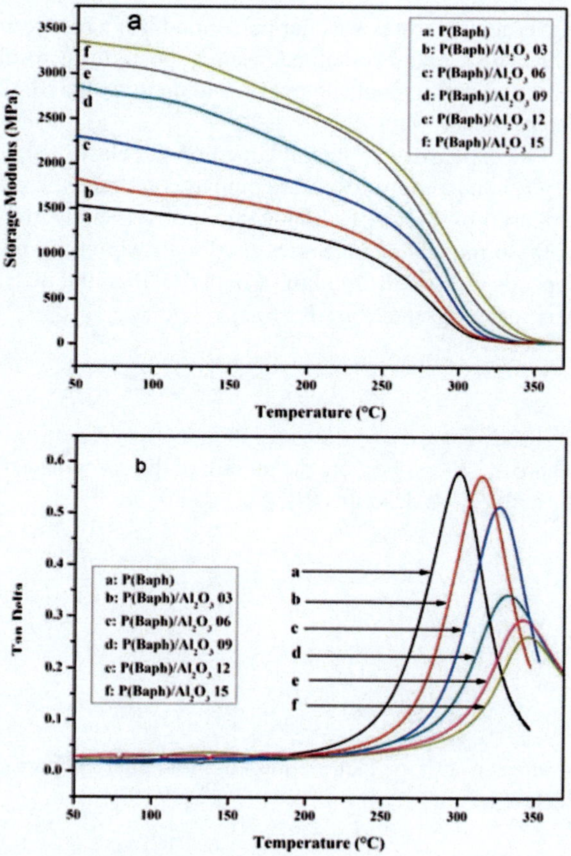

Fig. 8.1 DMA: storage modulus (**a**) and Tan delta (**b**) of the cured BAPh/Al_2O_3 nanocomposites (Derradji et al. 1549)

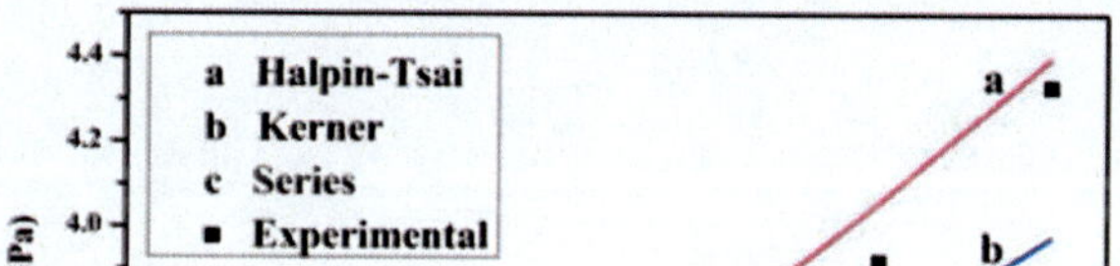

Fig. 8.2 Calculations and experimental measurements of the cured P(BAPh)/Al_2O_3 tensile modulus (Derradji et al. 1549)

higher temperatures was not performed. As a result, the nanocomposites with $T_g <$ 400 °C were mainly obtained, whereas post-curing resulted in achieving T_g of 446 °C for the neat phthalonitrile matrix and up to 540–570 °C for the nanocomposites (see Tables 6.3 and 7.1).

The experimental data obtained on the elastic modulus of the studied P(BAPh)/ Al_2O_3 nanocomposites were compared with the three theoretical models previously proposed by Series, Halpin–Tsai, and Kerner (Derradji et al. 1549). The simplest of these models is the Series model, in which the modulus of the nanocomposite depends only on the moduli of both the filler and matrix and the composition of the composite as seen from (8.1):

$$\frac{1}{E_c} = \frac{V_m}{E_m} + \frac{V_f}{E_f} \tag{8.1}$$

where E_c, E_f, and E_m are the moduli of the composite, filler, and matrix, respectively. V_f and V_m are described by (8.2) and (8.3):

$$E_c = \frac{1 + \xi\eta V_f}{1 - \eta V_f} \tag{8.2}$$

$$\eta = \frac{\frac{E_f}{E_m} - 1}{\frac{E_f}{E_m} + \xi} \tag{8.3}$$

where ξ is a shape factor equal to 2 for spherical particles. Kerner model is given by (8.4)–(8.7):

$$\frac{E_c}{E_m} = \frac{1 + ABV_f}{1 - B\varphi V_f} \tag{8.4}$$

$$A = \frac{7 - 5\vartheta}{8 - 10\vartheta} \tag{8.5}$$

$$B = \frac{\frac{E_f}{E_m} - 1}{\frac{E_f}{E_m} + A} \tag{8.6}$$

$$\varphi = 1 + \frac{1 - V_m}{V_m^2} \tag{8.7}$$

where A is a constant that depends on both the shape of the filler and the Poisson's ratio of the resin. B is also a constant related to the modulus of the filler and that of the matrix.

Figure 8.2 shows the experimental values of tensile modulus for the P(BAPh)/ Al_2O_3 nanocomposites and the modulus values predicted by the above three calculation models. It can be seen that as the nanoparticle content in the composites increases, the modulus increases from 2.9 GPa in the neat matrix to 4.3 MPa in the nanocomposite with 15 wt.% Al_2O_3. One can see that the best reproduction of the experimental values is given by the Halpin–Tsai model. The Kerner model also satisfactorily predicts nanocomposite modulus values on the whole and even gives a more accurate prediction for less filled matrices. At the same time, the simplified Series model clearly fails to describe the experimental data and is far from reality.

Finally, thermal stability of the neat resin and P(BAPh)/Al_2O_3 nanocomposites at various nano-Al_2O_3 contents under a nitrogen atmosphere was evaluated by TGA (Derradji et al. 1549). Especially illustrative were the temperatures at 5% weight loss ($T_{5\%}$) and 10% weight loss ($T_{10\%}$). So, the $T_{5\%}$ and $T_{10\%}$ of the unfilled matrix were 461 ° and 489 °C, respectively, whereas they increased up to 502 ° and 552 °C at 15 wt.% of the silanized nano-Al_2O_3 content.

Shan et al. (2017) synthesized and studied phthalonitrile nanocomposites with Al_2O_3 nanoparticles functionalized with nitrile groups on their surface. Compared with untreated nano-Al_2O_3, nitrile-functionalized Al_2O_3 (NC-Al_2O_3) particles (1 wt.%, 3 wt.%, or 5 wt.%) showed a more significant enhancement effect on the properties of the matrix. Storage modulus, glass transition temperature, flexural strength, and modulus of phthalonitrile matrix were improved with the incorporation of NC-Al_2O_3 nanoparticles. Nevertheless, on the whole, these nanocomposites were somewhat inferior in some properties to the above-described nanocomposites with silanized nano-Al_2O_3.

8.2 Phthalonitrile/Titania Nanocomposites

Derradji et al. (2016a) synthesized and studied a series of nanocomposites based on polymerized bisphenol-A phthalonitrile resin P(Baph) with different amounts (2 or 4, or 6 wt.%) of modified titania (TiO_2) nanoparticles. TiO_2 nanoparticles with an average size of about 60 nm were treated with the GX-540 (aminopropyltrimethoxysilane) coupling agent. To assure a good dispersion of the titania nanoparticles, the Baph/TiO_2 mixtures were vigorously stirred using a mechanical agitator for 4 h at the rate of 4000 rpm and then sonicated for 4 h. For the mechanical and thermal tests, these mixtures were cured by step heating at temperatures from 240° to 320 °C.

The structure of the nanocomposites was evaluated in Derradji et al. 2016a using FTIR, TEM, and XRD, and their properties were examined by DSC, DMA, and TGA. In addition, the microhardness and anticorrosive, protective properties of the nanocomposite coatings were defined. In the latter case, the measurements were performed by electrochemical impedance spectroscopy (EIS).

Figure 8.3 shows the FTIR spectra of native and treated TiO_2 nanoparticles.

The spectrum indicates effective silane grafting to the nanoparticle surface. Thus, peaks appearing at 2940 cm^{-1} and 2850 cm^{-1} were attributed to the asymmetric and symmetric CH_2 stretching vibrations, and the bands at 1455 cm^{-1}, 1305 cm^{-1}, and 998 cm^{-1} were assigned to the Si–O-C bonds, O-Si vibrations, and the Ti–O-Si vibrations, respectively.

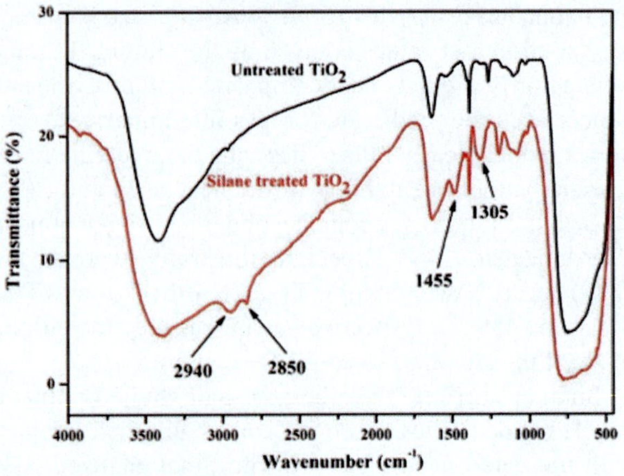

Fig. 8.3 FTIR spectra of native and treated TiO_2 nanoparticles (Derradji et al. 2016a)

The *X*-ray diffraction patterns of the nanocomposites with 2, 4, and 6% TiO_2 nanoparticles indicated that the amorphous structure of the matrix is retained in the nanocomposites, whereas the crystalline peaks of TiO_2 nanoparticles were detected in the diffractograms (Derradji et al. 2016a).

The DMA results obtained for the P(Baph)/TiO_2 nanocomposites are presented in Figs. 8.4a, b. The temperature dependencies of the dynamic (storage) modulus and tanδ show the following. At 50^{0}C, the storage modulus of the neat matrix was 1.83 GPa, and it increased sharply when TiO_2 nanoparticles were introduced, reaching 3.16 GPa at 6 wt.% nanoparticles. It is significantly that in this case the modulus increased approximately twice at 200 °C and three times at 300 °C. The shift of the relaxation peak of glass transition in nanocomposites led to an increase in T_g, estimated in the maxima of tanδ (T) peaks, from 314 °C to 364 °C. In addition, a significant decrease in the intensity of the glass transition relaxation peak was observed as titania nanoparticles were introduced (Fig. 8.4b).

In this work, as above for nanocomposites containing alumina nanoparticles, the experimental estimates of nanocomposite modulus were compared with the calculated values predicted from the Halpin–Tsai, Kerner, and Series models. It turned out that the experimentally obtained modulus values of nanocomposites with different contents of TiO_2 nanoparticles (from 2 to 6 wt.%) were in satisfactory agreement with the first two of these models (Derradji et al. 2016a).

Since the nanocomposites under consideration are intended for use as anticorrosive protective high-performance coatings, it was essential to evaluate changes in their microhardness due to the introduction of TiO_2 nanoparticles. As it can be seen from Fig. 8.4c, the microhardness of the neat matrix is 400 MPa and its value increases with increasing content of nanoparticles up to 500 MPa at 6 wt.% loading.

Undoubtedly, the enhanced thermomechanical properties of the nanocomposites under discussion and their increased microhardness could be explained by such factors as high hardness of titania nanoparticles, their surface silanization, and the ultrasound sonication of the mixture during nanocomposite fabrication. These factors

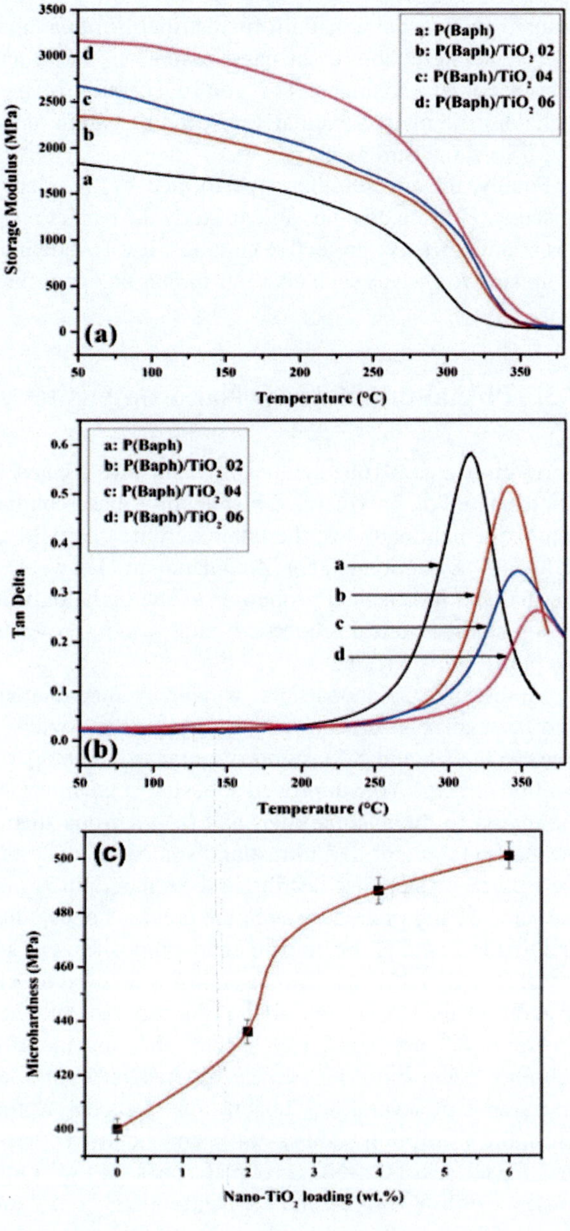

Fig. 8.4 Mechanical properties of the cured P(Baph)/TiO_2 nanocomposites. DMA: evolution of storage modulus (**a**) and Tan delta (**b**) versus temperature curves; (**c**) microhardness versus TiO_2 content dependence (Derradji et al. 2016a)

ensured good dispersion, uniform distribution of nanoparticles in the matrix, elevated interfacial interactions in the nanocomposites, and as a result the substantial suppression of matrix dynamics. The good dispersion of the titania nanoparticles in the phthalonitrile matrix, even at a maximum content of 6 wt. %, was confirmed in the TEM/SEM measurements.

Finally, the measurements performed by the electrochemical impedance spectroscopy (EIS) made it possible to study the barrier properties of these nanocomposites as anticorrosive protective coatings. It was shown that these nanocomposites are promising for use as such coatings in marine environments (Derradji et al. 2016a).

8.3 Phthalonitrile/ZnO Nanocomposites

Derradji et al. (2016b) synthesized and investigated high-performance nanocomposites based on polymerized bisphenol-A-based phthalonitrile resin (P(Baph)) and zinc oxide nanoparticles; the latter were treated with aminopropyltrimethoxysilane (GX-540) as coupling agent. ZnO nanoparticles were taken to improve not only the thermal and mechanical properties of the phthalonitrile matrix, but also (not least) to improve the optical properties, namely to improve the UV-shielding efficiency of the phthalonitrile resin.

Modified ZnO nanoparticles with an average diameter of 60 nm were introduced into the matrix in the amounts of 2 wt.% or 4 wt.%, or 6 wt.%. Again, as above (see Sects. 8.1 and 8.2), good dispersion of nanoparticles was achieved due to a twofold effect: (a) the improved adhesion at the interfaces owing to covalent binding the matrix to the nanoparticles and (b) vigorous stirring of the initial mixture in a mechanical agitator and ultrasound sonication. The P(Baph)/ZnO nanocomposites were cured by stepwise heating in the temperature range from 240 to 320 °C. Thus, the same curing procedure as in the previous ones with the preparation of alumina- and titania-based phthalonitrile nanocomposites was used (see above). FTIR, TEM, DMA, TGA, and X-ray diffraction test were used for characterizing these materials. In addition, the UV–visible absorption and transmittance spectra were also recorded.

Figure 8.5 presents FTIR spectra showing the effective grafting of the silane coupling agent onto the surface of ZnO nanoparticles. This is confirmed by the appearance of several new absorption bands due to the treatment of nanoparticles. Thus, the absorption bands at 3406 cm^{-1} and 1571 cm^{-1} refer to the stretching and bending modes of the -NH_2 group, the peaks at 1130 cm^{-1} and 1011 cm^{-1} indicate the presence of Si–O bonds, and the bands at 2937 cm^{-1} and 2886 cm^{-1} were attributed to the –CH_2 stretching vibrations.

TEM images of the P(Baph)/ZnO nanocomposites at the maximal nano-ZnO loading of 6 wt.% showed that a good state of dispersion has been achieved even at this nanofillers content. Some agglomerates of less than 200 nm in size were also present, i.e., they consisted of only two or three nanoparticles (Derradji et al. 2016b).

Figure 8.6 shows the tensile stress–strain curves obtained for the neat matrix and three nanocomposites. One can see a 1.5-fold increase in strength and an increase in

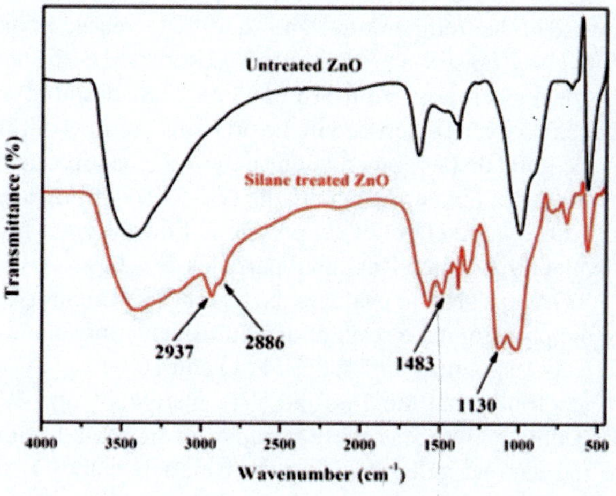

Fig. 8.5 Fourier transform infrared spectra of native and treated ZnO nanoparticles (Derradji et al. 2016b)

modulus of the latters, for example, from 2.9 GPa for the neat matrix to 4.3 GPa for the nanocomposite with 6 wt.% ZnO.

Figure 8.7 shows the results of the dynamic mechanical analysis of the cured P(Baph)/ZnO nanocomposites. As in cases where other metal oxide nanoparticles (Al_2O_3, TiO_2, see above) were introduced, significant effects of improving thermo-mechanical properties are also observed here, increasing as the content of nanoparticles in the composites increases. It can be seen that at 50 °C the dynamic (storage) modulus of the neat matrix equals 1.94 GPa, and for the composite with 6 wt.% nano-ZnO, it increases to 3.65 GPa. It is significant that high modulus values at the level of 3 GPa are retained by this nanocomposite up to temperatures around 270 °C, while for the neat matrix its value decreases sharply—to 0.5 GPa. In this case, as ZnO nanoparticles were added, the relaxation peak of glass transition, tanδ (T), shifted

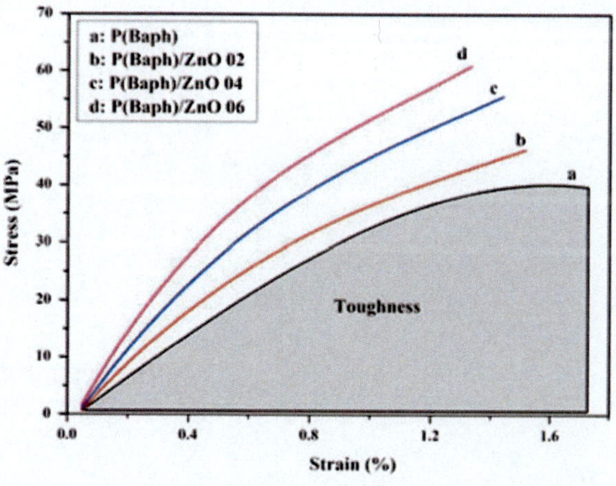

Fig. 8.6 Tensile stress–strain curves of the P(Baph)/ZnO nanocomposites (Derradji et al. 2016b)

toward higher temperatures and sharply decreased in intensity. As a result, at 6 wt.% ZnO glass transition temperature T_g, estimated as the temperature of the maximum of this peak, increased from 315 ° to 359 °C. Such improvements in the thermomechanical behavior, as discussed in the previous sections of this chapter, could be attributed to the good dispersion and adhesion of the nanoparticles to the matrix, owing to the treatment of the particles with the GX-540 coupling agent. These results also clearly indicate a large effect of suppression of the polymer matrix dynamics by embedded, covalently attached ZnO nanoparticles.

TGA was used to evaluate the effect of ZnO nanoparticles on the thermal stability of the phthalonitrile resin under a nitrogen atmosphere. Data related to the temperature corresponding to 5 wt.% ($T_{5\%}$) and 10 wt.% ($T_{10\%}$) weight loss were obtained. It was found that the $T_{5\%}$ and $T_{10\%}$ were 464° and 493 °C for the unfilled matrix, respectively, whereas these parameters increased with the increase of the amount of the nanoparticles reaching their highest values at 6 wt.% of nano-ZnO content when $T_{5\%}$ and $T_{10\%}$ increased by 33^0 and 61^0C, respectively. The enhancement of

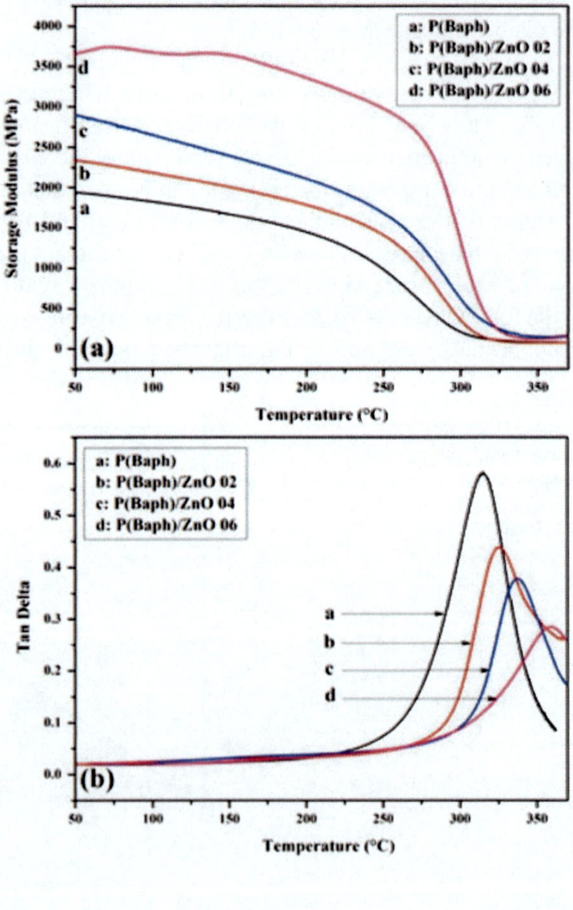

Fig. 8.7 Evolution of storage modulus (**a**) and Tan Delta (**b**) of the cured P(Baph)/ZnO nanocomposites (Derradji et al. 2016b)

the thermal stability of the examined nanocomposites in the inert medium authors (Derradji et al. 2016b) explained by the influence of two factors: the suppressive effect of nanoparticles on the mobility of the matrix and by the barrier effect.

Finally, in order to evaluate the nanocomposites under consideration as high-performance coating materials for strong UV protection applications, their optical properties were investigated. The UV–visible transmittance spectra of the P(Baph)/ZnO nanocomposites at different amounts of reinforcement showed the following: adding more nano-ZnO particles improved the UV-shielding properties of the neat phthalonitrile resin. For example, it turned out that the neat phthalonitrile matrix already exhibited a UV-shielding behavior of 78%, which confirms that it possesses good shielding properties. At the same time, a nanocomposite with 6 wt% nano-ZnO showed a UV-shielding behavior of 86.5%. This increase in the absorption capacity was attributed to the special optical properties of ZnO nanoparticles.

References

M. Derradji, N. Ramdani, T. Zhang, J. Wang, L.-D. Gong, X.-D. Xu et al., Thermal and mechanical properties enhancements obtained by reinforcing a Bisphenol-A based Phthalonitrile resin with silane surface-modified alumina nanoparticles. Polym. Compos. **38**, 1549 (2017). https://doi.org/10.1002/pc.23722

M. Derradji, N. Ramdani, T. Zhang, J. Wang, l.-D. Gong, X.-D. Xu, et al., Effect of silane surface modified titania nanoparticles on the thermal, mechanical, and corrosion protective properties of a bisphenol-a based phthalonitrile resin. Prog. Org. Coat., **90**, 34 (2016a). https://doi.org/10.1016/j.porgcoat.2015.09.021

M. Derradji, N. Ramdani, L-d Gong, J. Wang, X.-D. Xu, Z.-W. Lin, et al., Mechanical, thermal, and UV-shielding behavior of silane surface modified ZnO reinforced phthalonitrile nanocomposites. Polym. Adv. Technol., **27**(7), 882 (2016b). https://doi.org/10.1002/pat.3744

S. Shan, X. Chen, Z. Xi, X. Yu, X. Qu, Q. Zhang, The effect of nitrile functionalized nano-aluminum oxide on the thermomechanical properties and toughness of phthalonitrile resin. High Perform. Polym. **29**(1), 113 (2017). https://doi.org/10.1177/0954008316631593

Chapter 9
Other Types of Phthalonitrile Nanocomposites

Abstract The structure and properties of six types of cured PhN nanocomposites containing functionalized silicon nitride, boron nitride, MAX phase, tungsten nanoparticles, MXene nanosheets, and graphite nanoplatelets are described. In all cases, the introduction of nanoparticles led to a significant improvement in the thermomechanical properties of the matrix. The most striking examples of their positive influence are as follows. For PhN/BN nanocomposites, the storage modulus increased from 1.94 GPa up to 7.2 GPa at 50 °C and from ~ 0 to 4 GPa at 300 °C. The flexural modulus and strength at 20 °C increased from 1.8 GPa and 75 MPa to 8.5 GPa and 190 MPa, respectively, and a 26-fold increase in thermal conductivity was registered. The nano-MXene phase (3 wt.%) resulted in increasing mass loss temperature $T_{10\%}$ from 493 °C to 649 °C for the nanocomposite. For nanotungsten-containing composites, the effectiveness in shielding against gamma radiation was shown. Adding graphite nanoplatelets decreased the electric resistivity of the matrix from ~10^{14} Ω cm down to 10^7 Ω cm. The data of X-ray photoelectron spectroscopy, IR spectroscopy, and other information allowed us to conclude that the main reasons for the ultra-high thermal properties of phthalonitrile nanocomposites are the leading role of uniquely heat-stable phthalocyanine heterocycles in their structure and constraining matrix dynamics by the covalently embedded nanoparticles.

9.1 Phthalonitrile/Silicon Nitride Nanocomposites

Derradji et al. (2015) developed and studied a new kind of nanocomposites based on bisphenol-A-based phthalonitrile resin reinforced with silicon nitride (SiN) nanoparticles. The nanoparticles were treated by the silane coupling agent GX-540 (aminopropyltrimethoxysilane). 2,2-bis [4-(3,4-dicyanophenoxy)phenyl] propane (BAPh) and 4-Aminophenoxy phthalonitrile (Apph, curing agent) monomers were mixed at a weight ratio of 90:10, and then the proper amount of the treated nanoparticles of silicon nitride was added in different amounts ranging from 3 wt.% to 15 wt.%. The SiN nanoparticles were in the form of powder with a crystalline structure having a density of 3.44 g/cm^3 and an average diameter of particles of 50 nm. The mixtures

V. A. Bershtein and P. N. Yakushev, *High-Temperature Polymer Nanocomposites Based on Heterocyclic Networks from Nitrile Monomers*, Springer Series in Materials Science 334, https://doi.org/10.1007/978-3-031-32943-2_9

were vigorously stirred, and the samples were cured by the hot compression molding technique at 240–300 °C. The cured nanocomposites were noted as P(BAPh)/SiN ones.

These nanocomposites were characterized by FTIR, DSC, DMA, TGA, and SEM techniques. The main attention was paid to the analysis of the thermomechanical properties of these materials.

First of all, the successful modification of the nanoparticle surface by the silane coupling agent was confirmed by FTIR spectra. Furthermore, these spectra proved that the Apph curing agent has effectively cured the BAPh monomers, and, as a result, the heterocyclic matrix network has been formed.

Figure 9.1 presents the DMA data obtained for the neat matrix and five cured P(BAPh)/SiN nanocomposites containing 3, 6, 9, 12, and 15 wt,% nanofillers. One can see a strong improvement in the thermomechanical properties of the polymer matrix due to the introduction of SiN nanoparticles, and the positive effect increases with increasing the concentration of nano-SiN particles.

The storage modulus of the neat matrix is equal to 1.7 GPa at 50 °C, and at the maximum fillers loading it reaches 4 GPa. The high level of modulus values, more than 2.5 GPa, is also maintained for this nanocomposite at high temperatures, up to 320–330 °C when the modulus of the neat matrix becomes negligibly small. The T_g of the cured unfilled matrix, obtained from the maximum of a *tanδ* peak, was 300 °C. *Tanδ* as a function of temperature shows that with the introduction and increase in the concentration of SiN nanoparticles in the composite, the glass transition peaks shift toward higher temperatures, and the height of the peaks decreases significantly. A value of $T_g = 360$ °C has been recorded for the cured nanocomposite when the amount of the SiN nanoparticles reached 15 wt.%.

The reasons for such a significant improvement in the thermomechanical properties of the matrix in the presence of SiN nanoparticles are quite clear. It is (a) satisfactory dispersion of nanoparticles in the matrix; (b) good adhesion of the matrix and nanoparticles due to the treatment of the latter with the GX-540 silane coupling agent, i.e., actually chemical hybridization of these components because of the reaction between amine and nitrile atomic groups, and (c) significant suppression of the matrix dynamics by nanoparticles. The results of the performed SEM analysis (Derradji et al. 2015) did not contradict these assumptions.

TGA of the neat matrix and a number of the nanocomposites under consideration also revealed increased thermal stability of the latters in the air environment. In doing so, in particular, the degradation temperatures at 5% weight loss ($T_{5\%}$) and 10% weight loss ($T_{10\%}$) were evaluated. It turned out that the $T_{5\%}$ and $T_{10\%}$ of the unfilled matrix were 459° and 488 °C, respectively, whereas for the nanocomposite with 15 wt.% of SiN nanoparticles $T_{5\%}$ and $T_{10\%}$ increased by 37° and 58 °C respectively. This effect could be associated with the abovementioned reasons and also related to the shielding effect of the nano-SiN particles that act as a barrier. Thus, silicon nitride nanoparticles treated with silane coupling agent GX-540 have a definite positive effect on thermomechanical properties, T_g, and thermal stability of the phthalonitrile matrix.

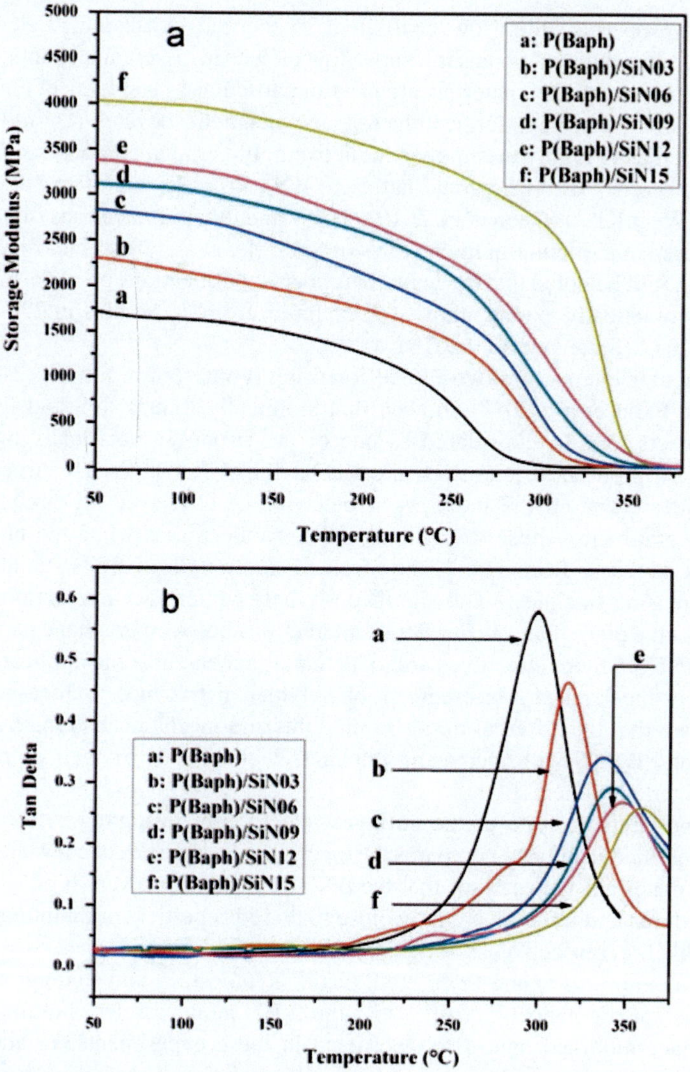

Fig. 9.1 DMA, 1 Hz: evolution of storage modulus (**a**) and tanδ (**b**) of the cured BAPh/SiN nanocomposites (Derradji et al. 2015)

9.2 Phthalonitrile/Boron Nitride Nanocomposites

In the last decade, there has been a great deal of interest in the production and numerous applications of boron nitride nanoparticles (nano-BN), including BN nanosheets (BNNS), and their nanohybrids because of their unique physical properties. That is explained by the unusual combination of their properties, such as low specific density, high thermal stability, oxidation resistance, exclusively high mechanical properties, ultra-low coefficient of thermal expansion, high thermal conductivity, low dielectric constant, super-low dielectric losses, and chemical inertness. Therefore, nano-BN materials are used in particular as elements of electronic, plasmonic, optoelectronic, semiconductor, and magnetic devices (Derradji et al. 2017), as well as in highly thermally conductive polymer nanocomposites (Yu et al. 2018) since the BN with hexagonal lattice (h-BN) possesses high thermal conductivity (240 $W(m.K)^{-1}$) (Chen et al. 2019). These and other applications of nano-BN are considered in particular in the reviews of Shtansky et al. (2018) and Zhang et al. (2017). It should be noted that the annual number of publications on nano-BN materials grows practically exponentially, for example, from 25 articles in 2007 to 250 articles in 2017 (Shtansky et al. 2018).

Of particular interest are two-dimensional hexagonal boron nitride (2-D-hBN) nanosheets. Falin et al. (2017) showed that atomically thin BNNS and few-layer BN nanosheets may be considered as one of the strongest electrically insulating materials. Of importance, the mechanical behavior of few-layer BN turned out to be quite different from the mechanical behavior of few-layer 2-D graphene. In contrast to graphene, whose strength considerably decreases when the number of nanolayers increases from 1 to 8, the mechanical strength of BNNS is not sensitive to increasing thickness. The significantly better interlayer interactions of BN nanosheets take place than van der Waals interactions between graphene nanolayers; it makes BNNS a more attractive candidate than graphene in some applications, for example, for mechanical reinforcement of polymer matrix in the nanocomposites. It was shown that the different trends in modulus and mechanical strength between graphene and BNNS with increasing thickness were caused by their dramatically different interlayer interactions.

BN nanosheets are one of the strongest electrically insulating materials. The scheme presented in Fig. 9.2 compares the mechanical properties of a few electrically insulating materials. One can see that the BN nanosheets have very high values of a Young modulus and fracture strength, close to these properties for diamond: ca. 860 GPa and 70 GPa, respectively.

Derradji et al. (2017) and Chen et al. (2019) synthesized and studied the highly filled phthalonitrile nanocomposites with the h-BN nanoparticles. Both native and silane surface-modified nano-BN were used in these experiments. In addition to evaluating the structure and thermomechanical properties of these nanocomposites, the main goal of the study was to increase the thermal conductivity of the materials. The purpose of introducing increased amounts of nanofiller was to overcome the phonon scattering problem which is the main cause of low thermal conductivity in polymeric materials. At high content of nanoparticles in the matrix, the latter tend to touch each other forming "conductive paths" and therefore reducing the thermal resistance of the nanocomposite.

In the study (Derradji et al. 2017), a high-performance bisphenol-A-based phthalonitrile (Baph) resin was chosen as the matrix for preparing the nanocomposites with different contents of the native and GX-540 silane coupling agent-modified h-BN nanoparticles, ranging between 5 wt.%. and 30 wt.%. The BN nanoparticles had an average diameter of about 60 nm. Before curing, the mixtures were mechanically

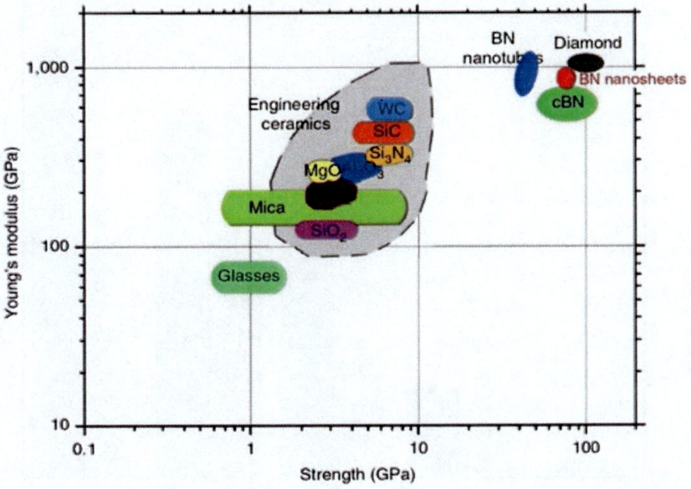

Fig. 9.2 Modulus-strength graph. The mechanical properties of different electrically insulating materials, including monolayer and few-layer BN, are compared (Falin et al. 2017)

stirred and sonicated. Samples of the neat matrix and the subsequent nanocomposites were cured by hot pressing at 240–320 °C. TEM, SEM, FTIR, TGA analyses, and mechanical and thermal conductivity measurements have been performed.

Derradji et al. (2017) obtained a number of very impressive results from the introduction of the BN nanoparticles into the matrix, which could be expected due to the exceptionally high properties of this type of nanoparticles.

It should first be mentioned that FTIR analysis confirmed the effectiveness of the chemical treatment (modification) of the surface of BN nanoparticles by a silane coupling agent. Thus, the spectrum of the treated BN nanoparticles revealed the presence of new absorption bands, for example, the bands at 2938 and 2840 cm^{-1} of the -CH_2 vibrations and the absorption near 1100 cm^{-1}, related to the Si–O vibration. Consequently, there was the grafting of silane to BN.

Further, TEM analysis showed that the surface modifying treatment of BN nanoparticles had a distinctly positive effect on the quality of their dispersion even at 20 and 30 wt.% nano-BN loadings. Additionally, the nanofillers tend to form "a conductive network" favorable for the heat transfer.

Phthalonitrile polymer as any other polymeric material is characterized with a poor thermal dissipation affecting the performance, lifetime, and reliability of PhN-based electronic devices. Hence, developing thermally conductive PhN materials is of primary importance. Therefore, the effective way to enhance the thermal conductivity of polymeric materials is to prepare composites by introducing thermally conductive inorganic particles.

Figure 9.3 presents all the thermal conductivity results obtained for the investigated materials, including the neat matrix and nanocomposites with different BN nanoparticle content and surface state. One can see that the neat matrix exhibited a

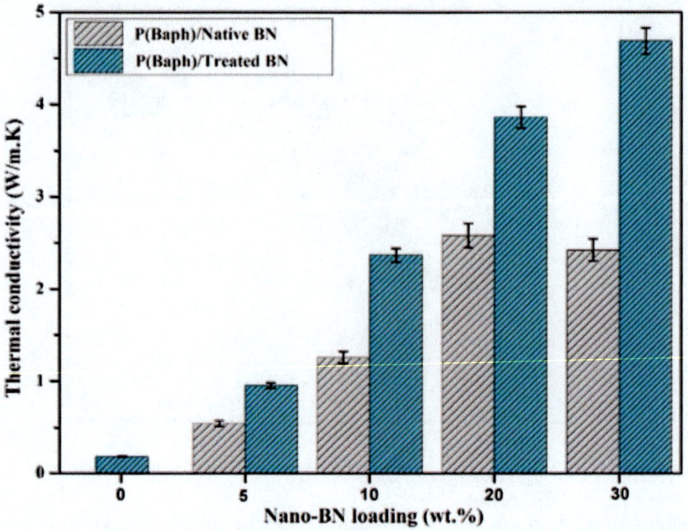

Fig. 9.3 Thermal conductivity of the phthalonitrile resin reinforced with different amounts of native and treated nano-BN (Derradji et al. 2017)

thermal conductivity value only of 0.18 W/m K. At the same time, the thermal conductivity increased sharply and systematically with the introduction and increasing concentration of nano-BN. The reason for this is that the BN with hexagonal lattice (h-BN) possesses very high thermal conductivity (240 W/m K) (Chen et al. 2019). The native, untreated BN nanoparticles enhanced the thermal conductivity of the neat matrix, however, in the case of nanoparticles with a modified surface the effects were significantly higher. In the limiting case of the nanocomposite with 30 wt.% treated nano-BN, the thermal conductivity increased to 4.69 W/m K, i.e., by a factor of 26. The lower thermal conductivity values obtained with the native BN were ascribed to the high interfacial thermal resistance between the matrix and the nanofillers.

Figure 9.4 shows the DMA results, temperature dependencies of Tanδ (relaxation peaks) and storage modulus obtained for the neat matrix and a number of the cured P(Baph)/BN nanocomposites containing 5 wt.%, 10 wt.%, 20 wt.%, and 30 wt.% BN nanoparticles.

One can see that the stiffness, designated as the storage modulus value at 50 °C, and the T_g determined from the maximum of the Tanδ peaks were equal for the neat matrix to 1.94 GPa and 315 °C, respectively. The introduction of nano-BN resulted in significant changes in these characteristics. Especially impressive is the multiple growth of the storage modulus at 50 °C, up to 7.2 GPa, at a BN content of 30 wt.% in the nanocomposite. Quite high storage modulus values are retained by this nanocomposite even at temperatures of 300–320 °C when the storage modulus of the neat matrix becomes extremely low. Moreover, at the maximum nano-BN content, the T_g of the nanocomposite was found to be 47 °C higher than that of the neat matrix.

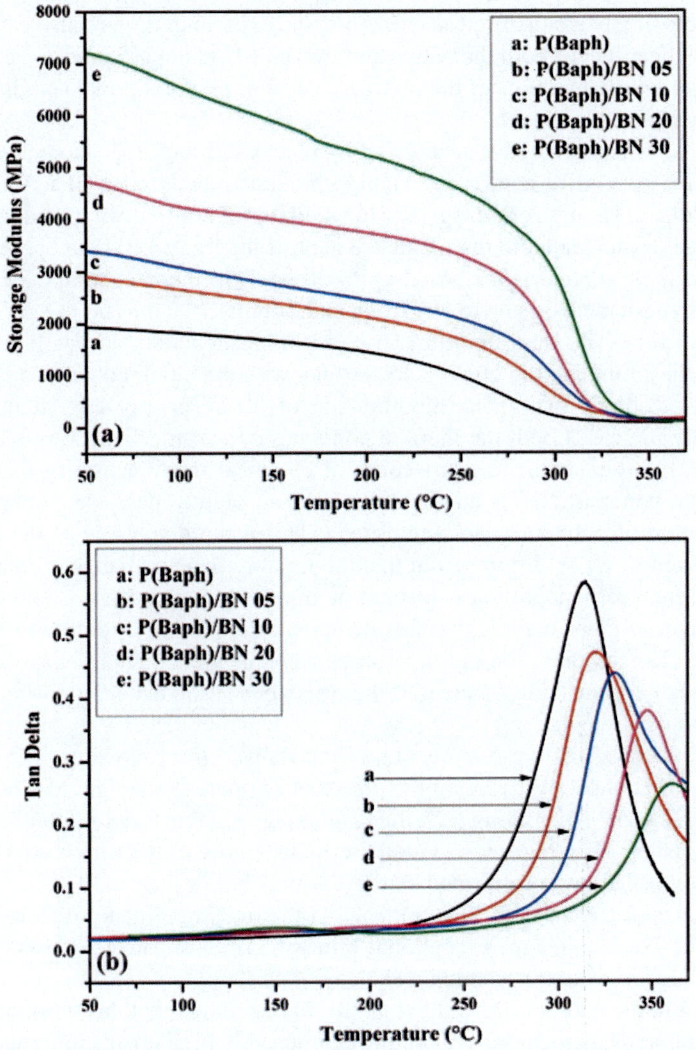

Fig. 9.4 Evolution of storage modulus (**a**) and Tan delta (**b**) of the cured P(Baph)/BN nanocomposites with 5, 10, 20, and 30 wt.% nanofillers (Derradji et al. 2017)

Of course, the extraordinary increase in stiffness of BN-containing nanocomposites is explained by the extra-high, "diamond-like" modulus of BN nanoparticles. The suppression of the matrix dynamics by the attached modified BN nanoparticles also contributes here to a certain extent. The latter is confirmed not only by the increase in T_g, but also by a significant decrease in the height of the relaxation glass transition peaks in the nanocomposites (Fig. 9.4b).

The aim of the work (Derradji et al. 2017) was to simultaneously enhance both thermal conductivity and thermomechanical properties of the phthalonitrile matrix with the help of BN nanoparticles. Therefore, the experiments were also carried out to evaluate the effect of both the native and treated BN nanoparticles on the flexural strength and Young modulus of the matrix, as well as on the thermal stability of the nanocomposites under study.

Figure 9.5 shows that the neat P(Baph) matrix exhibits low flexural strength (75 MPa) due to its brittle nature, and Young's modulus equals to 1.8 GPa. At the same time, the flexural strength and especially the modulus of elasticity of nanocomposites can increase dramatically, to the greatest extent, if the treated BN nanoparticles are used and if their content is increased up to 30 wt.% in the nanocomposite. In this case, the strength increases up to 190 MPa, and Young's modulus up to 8.5 GPa, i.e., almost by 5 times. At the same time, if nanocomposites are made with the native nano-BN, the strengthening effect is lower than with the modified nanoparticles.

Thus, it is shown that silane modification of the surface of BN nanoparticles significantly improves both the thermal conductivity and mechanical properties of the matrix. In other words, the procedure of chemical modification of the surface of inorganic nanoparticles is an important and absolutely necessary operation in the fabrication of polymer nanocomposites. The fractured surfaces of the samples were analyzed by SEM. It turned out that the fracture pattern changed dramatically in the presence of nanoparticles. Instead of the development of a single crack in a brittle neat polymer matrix, the nanocomposites exhibited a multitude of small cracks that changed direction multiple times when they encountered nanoparticles. This eventually dramatically increased the resistance to the nanocomposite fracture process and its strength.

Finally, the TGA of these nanocomposites showed the following. The starting decomposition temperatures for all the prepared samples at a weight loss of 5 wt.% ($T_{5\%}$) and 10 wt.%($T_{10\%}$) along with the residual weight (char yield) after heating to 800 °C (Y_c) have been registered. Good thermal stability of the neat matrix under a nitrogen atmosphere was confirmed by the $T_{5\%}$ and $T_{10\%}$ values of 464° and 493 °C, respectively, and a char yield $Y_c = 69.8\%$. At the maximum nanofillers loading of 30 wt.%, the $T_{5\%}$ reached the exceptional value of 551 °C and no $T_{10\%}$ was observed since the char yield at 800 °C remained $Y_c = 92.2\%$.

As a result, the authors (Derradji et al. 2017) concluded that this work on highly filled P(Baph)/BN nanocomposites achieved unusually high results in terms of high thermal conductivity, mechanical properties, and thermal stability. These materials can be used, therefore, in many applications requiring the combining of good thermal conductivity with excellent thermal and mechanical properties of the materials.

9.3 Phthalonitrile/MAX Phase Ceramics Nanocomposites

Of considerable interest are the recent works of Derraji et al. (Derradji et al. 2018a, 2019; Henniche et al. 2018), which describe the synthesis and properties of phthalonitrile nanocomposites with a rather high content of ceramic phase nanoparticles—the so-called MAX phase (Ti_3SiC_2 (Derradji et al. 2018a) or Ti_3AlC_2 nanoparticles

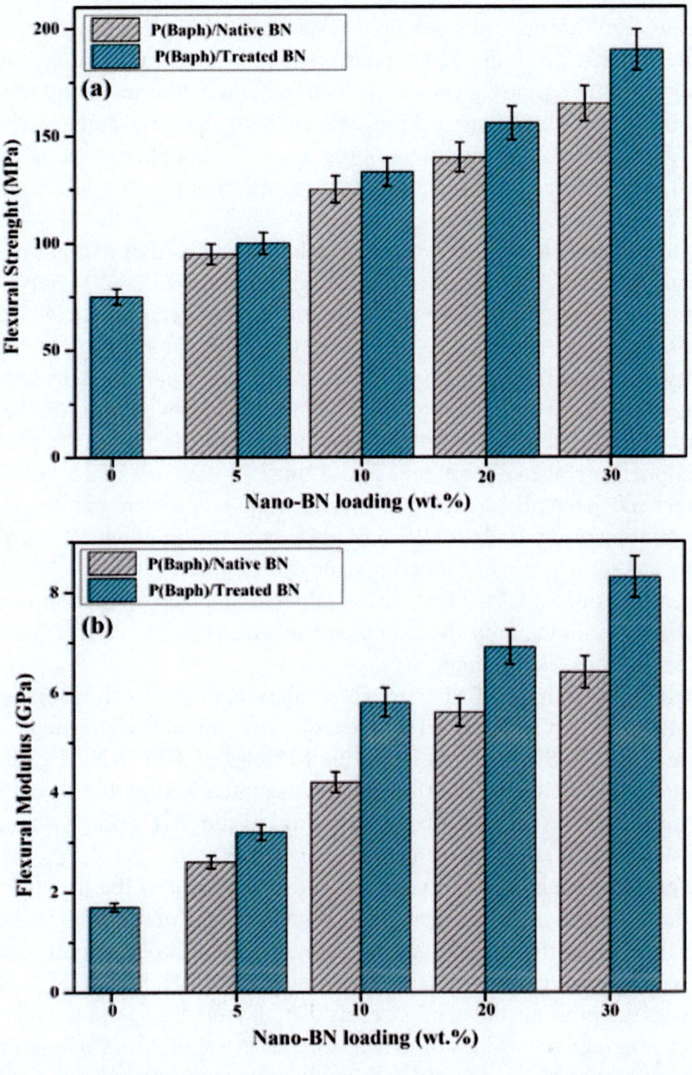

Fig. 9.5 Flexural strength (**a**) and modulus (**b**) of the phthalonirile resin and its nanocomposites with native and treated BN (Derradji et al. 2017)

(Henniche et al. 2018)) as well as "MXene phase" ($Ti_3C_2(OH)_2$ nanosheets) (Derradji et al. 2019) (see Sect. 9.4). Usually, the main focus of polymer nanocomposites is on such important factors as improved dispersion of nanoparticles in the matrix, adhesion of the matrix to nanoparticles, and the influence of the latter on the matrix dynamics. However, less attention is paid to the reinforcing phase itself. At the same time, new generations of perspective ceramics have been developed in recent years.

The Ti_3AlC_2 and Ti_2AlC compounds belong to the MAX (M_nAX_n) group of materials where $n = 1$, 2, or 3, M is a transitional metal, A is Al or Si, and X is C or N (Barsoum 2000). Their dual behavior resulting from the combination of metals and ceramics is what gives the MAX phase materials their outstanding properties. These features include good thermal and electrical conductivities, good mechanical properties, low density, excellent wear resistance, high hardness, and good oxidation resistance. Owing to this combination of properties, MAX phase particles have been widely used as reinforcement in metal and ceramic composites (Barsoum 2000; Eklund et al. 2010).

In the work (Henniche et al. 2018), spherical-shaped Ti_3AlC_2 nanoparticles were prepared and subjected to surface modification using the GX-540 coupling agent (aminopropyl triethoxysilane). The surface modification of nanoparticles was aimed at improving their dispersity and adhesion to the matrix due to the hybridization reaction between the amine groups of the coupling agent and the nitrile groups of the forming matrix network. These modified nanoparticles were introduced into the phthalonitrile matrix (the bisphenol-A-based phthalonitrile monomer (2,2- bis[4-(3,4-dicyanophenoxy)phenyl]propane) in varying amounts up to 15 vol.%. These mixtures were mechanically stirred and sonicated, and then were cured, in the presence of 3-aminophenoxy phthalonitrile (Apph) as a curing agent, at temperatures from 240° to 320 °C. The cured nanocomposites are denoted below as P(Baph)/MAX or P(BAPh)/MAX ones. SEM, TEM, XRD, DMA, TGA, and an Instron instrument, as well as thermal conductivity measurements allowed to characterize the structure and properties of these nanocomposites.

As could be seen from the SEM images (Henniche et al. 2018), the average particle size of the MAX (Ti_3AlC_2) phase after mechanical treatment (ball milling) was 100–300 nm. The results of SEM estimation of the particles of the MAX (Ti_3SiC_2) phase were obtained in Derradji et al. 2018a and are presented in Figs. 9.6 and 9.7. In this case, one can see the broad statistical distribution of the MAX phase nanoparticles size, from ~1 nm to 500 nm, with an average size of 60 nm.

In the work (Henniche et al. 2018), the authors measured the thermal conductivity and thermomechanical properties of both the neat matrix and a number of nanocomposites with different MAX phase contents. The experimental results were compared with the predictions obtained from the Series, Halpin–Tsai, and Kerner models. Figure 9.8 presents the results of this comparison of experimental and calculated thermal conductivity values of the considered P(Baph)/MAX nanocomposites, for the cases of low and high nanofiller contents in the nanocomposites. One can see the following.

The thermal conductivity of the neat matrix is 0.18 W/m K, and it increases steadily with increasing MAX phase content, reaching a value of 0.55 W/m K at 15 vol.% nano-MAX. At low nanofiller content, the predictions can be considered acceptable. However, this is not the case at sufficiently high nanoparticle concentrations. In this case, the nanoparticles begin to touch each other, forming in fact, the "conductive paths", dramatically facilitating the transport of heat. The formation of such paths was confirmed by a high-resolution TEM. At the same time, TEM images indicated good dispersity of nanoparticles and the absence of formation of their microscale aggregates even at the maximal MAX nanofillers loading. It is clear that, under these conditions, the calculated models cannot predict the thermal conductivity values.

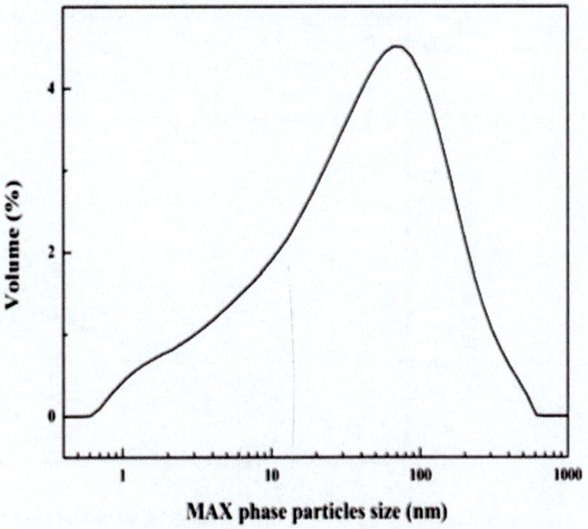

Fig. 9.6 Statistical distribution of the MAX phase (Ti_3SiC_2) nanoparticles size (Derradji et al. 2018a)

Fig. 9.7 SEM image of the as-prepared MAX phase (Ti_3SiC_2) nanoparticles after ball milling (Derradji et al. 2018a)

In this work (Henniche et al. 2018), the tensile properties including the strength, strain, modulus, toughness, thermomechanical behavior as well as thermal stability were also investigated for the neat matrix and P(Baph)/MAX nanocomposites.

As can be seen from Fig. 9.9, the introduction of the MAX phase affects the stress–strain curves otherwise than in the case of using traditional ceramics for reinforcement.

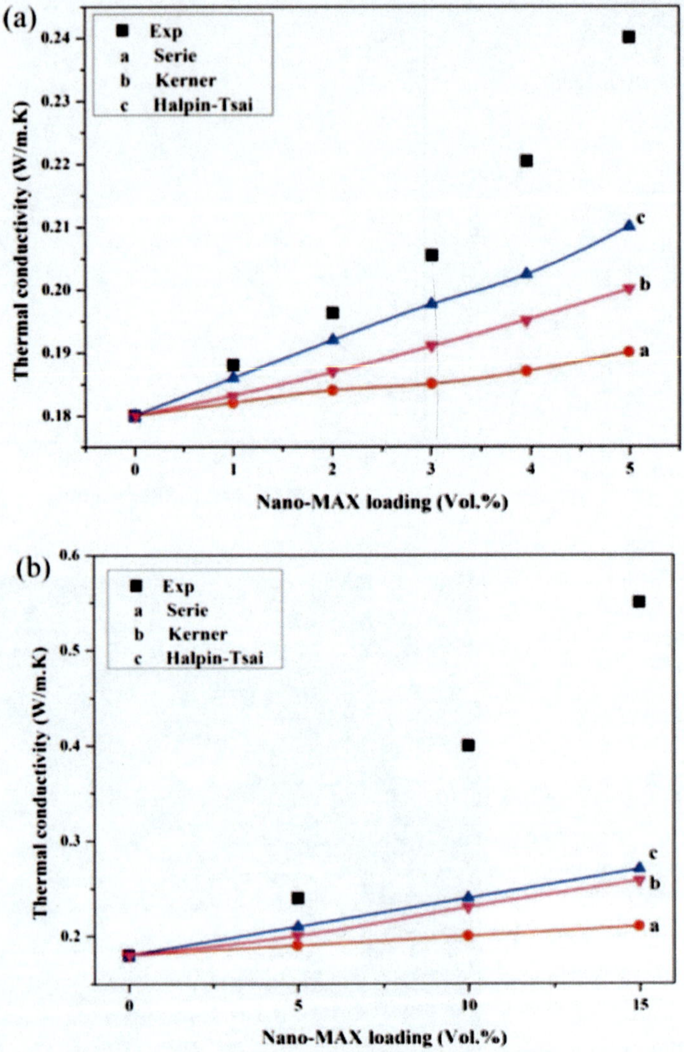

Fig. 9.8 Thermal conductivity calculations and experimental measurements for the P(Baph)/ Ti_3AlC_2 MAX nanocomposites at low (**a**) and high (**b**) nanofiller amounts (Henniche et al. 2018)

Really, with the increasing amount of introduced MAX phase, the tensile strength of the composite systematically increases, reaching the value of 88 MPa at 15 vol.% nano-MAX (more than twofold increase). However, the fracture strain remains the same as in the case of a neat matrix (about 1.7%). The point is that MAX phases are ductile or "soft" ceramics, and therefore their addition to rather brittle matrices, such as phthalonitrile, has no effect on ductility.

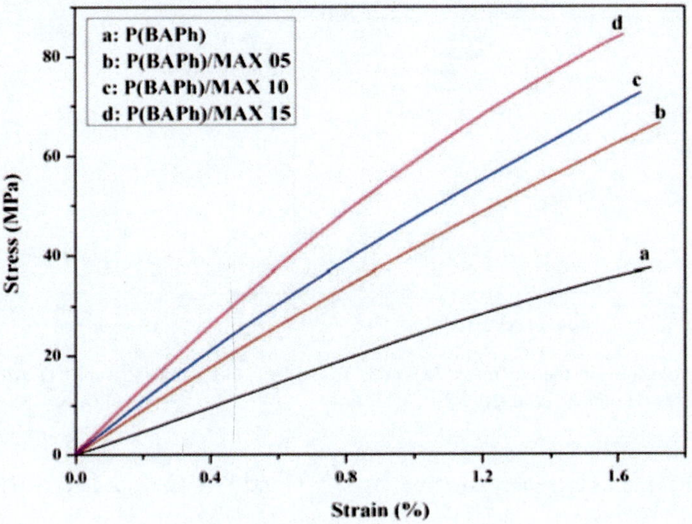

Fig. 9.9 Stress–strain curves of the cured BAPh/Ti_3AlC_2 MAX nanocomposites (Henniche et al. 2018)

The toughness of all the prepared materials was calculated by integrating the areas under the stress–strain curves. As expected, the "soft" MAX ceramics increased the toughness of the neat resin by 88%, 118%, and 129% for the MAX phase contents of 5, 10, and 15 vol.%, respectively (Henniche et al. 2018).

The DMA results of the nanocomposites under consideration, including the temperature dependencies of storage modulus and tan delta, are shown in Fig. 9.10. A strong reinforcing influence of the MAX phase on the thermomechanical behavior of the cured P(Baph) matrix can be seen. For the neat cured matrix, the storage modulus at 50 °C and the T_g value determined from the maximum of tan delta peaks equaled 1.7 GPa and 310 °C, respectively. These two characteristics increased with the amount of MAX phase introduced. As a result, at the maximum MAX phase content of 15% the storage modulus at 50 °C reached 4.2 GPa, and $T_g = 359$ °C. We emphasize that the significant increase in the stiffness of these nanocomposites was largely determined by the large MAX phase modulus (about 300 GPa), and the increase in T_g by the suppression of matrix dynamics. It should also be noted that the sharply increased modulus values are retained in the nanocomposite up to 270–280 °C, and the modulus of the latter at 300 °C is practically equal to that of the neat matrix at 50 °C (Fig. 9.10a).

Finally, TGA has been applied for a comparative evaluation of the effect of the MAX phase on the thermal stability of the P(baph) matrix. For this purpose, the starting decomposition temperatures at 5 wt.% ($T_{5\%}$) and 10 wt.% ($T_{10\%}$) mass loss were determined. It was found that in the case of neat matrix $T_{5\%}$ and $T_{10\%}$ were 464 and 493 °C, respectively, and at the highest MAX phase content in the nanocomposite,

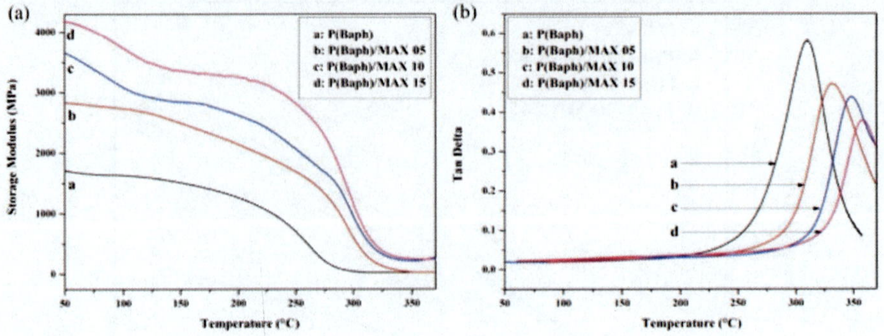

Fig. 9.10 Evolution of the stiffness (**a**) and T_g (**b**) of the cured P(Baph)/Ti_3AlC_2 MAX nanocomposites (Henniche et al. 2018)

these characteristics increased to values of 512 °C and 576 °C, respectively (Henniche et al. 2018).

9.4 MXene ($Ti_3C_2(OH)_2$) Nanosheet-Reinforced Phthalonitrile Nanocomposites

Recently, Derraji et al. (2019) published a paper on phthalonitrile nanocomposites in which a new generation of ceramics, graphene-like 2-D nanolayers, known as the MXene ($Ti_3C_2(OH)_2$), was used as a reinforcing phase. Due to their unique set of properties, MXenes are regarded as very promising additives for improving the mechanical, tribological, and other properties of polymers (Zhang et al. 2016; Taloub et al. 2019). The aim of the work (Derradji et al. 2019) was to obtain phthalonitrile-ceramics nanocomposites with excellent mechanical and thermal properties for extremely exigent applications.

MXene could be obtained in Derradji et al. (2019) (as in the article (Henniche et al. 2018)) from the precursors of the MAX phase (the Ti_3AlC_2 powder) by etching in HF acid. A schematic description of the synthesis process of the studied MXene nanosheets is given in Fig. 9.11. As with the MAX phase, the MXene nanolayers were modified with a silane coupling agent. A statistical study of the nanosheets' dimensions was performed, and the obtained results suggested an overall nanosheet size of about 60 nm. As in a number of studies discussed above, the P(BAPh) matrix was used to produce nanocomposites with MXene. The MXene ($Ti_3C_2(OH)_2$) nanosheets were introduced into the matrix in amounts of 1 wt.%, 2 wt.%, or 3 wt.%. The samples were cured by heating at 240–300 °C. SEM, TEM, TGA, DSC, and tensile tests were used to control the curing process and to evaluate the properties of the obtained materials.

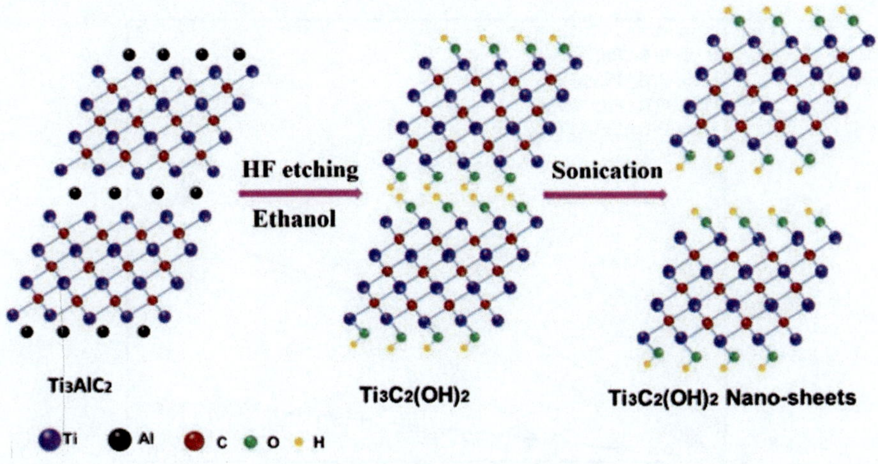

Fig. 9.11 Schematic description of the synthesis process of the studied MXene nanosheets (Derradji et al. 2019)

The tensile fractured surfaces of the P(BAPh)/MXene nanocomposites were analyzed by SEM, and the fracture character of nanocomposites, the uniform distribution of microcracks gave grounds to assume the acceptable dispersion of MXene nanolayers and their adhesion to the matrix. TEM analysis confirmed that a good state of dispersion was achieved even at 3 wt.% of nanoloading. However, at further increasing of MXene content, the presence of large-size agglomerates was observed which negatively affected the mechanical performance of the nanocomposites.

The tensile stress–strain curves shown in Fig. 9.12 indicate a gradual and strong increase in strength as the number of introduced MXene nanolayers increases. For example, the tensile strength of the neat matrix is about 40 MPa, whereas at 3 wt.% of nanoloading, the strength of the nanocomposite increases to 276 MPa. It is essential that the value of strain at rupture remains unchanged. The latter effect, as indicated above, is also characteristic of nanocomposites with MAX phase, i.e., it is inherent to the composites containing nanoparticles of this new class of "soft" ceramics.

Figure 9.13 and Table 9.1 show that the introduced MXene nanosheets effectively improved the thermal resistance of the P(BAPh) matrix. One can see that the starting decomposition temperatures, $T_{5\%}$ and $T_{10\%}$, and the char yield Y_c at 800 °C considerably increased in the nanocomposites increasing with the content of the MXene nanosheets. For instance, $T_{10\%}$ increased by 156 °C and Y_c grew from 69.8% to 87.8% when heating in a nitrogen atmosphere at the content of nanofillers of 3 wt.%. in the nanocomposite.

Thus, a distinctive, extraordinary feature of this new type of nanocomposites should be recognized as follows: with the introduction into the matrix of only 3 wt.% nanolayers of $Ti_3C_2(OH)_2$ as the 2-D layered structure of the MXene, the strength of nanocomposite increases by almost an order of magnitude—from 40 to 276 MPa at unchanged tensile strain at break, and the temperature of 10% mass loss when heated

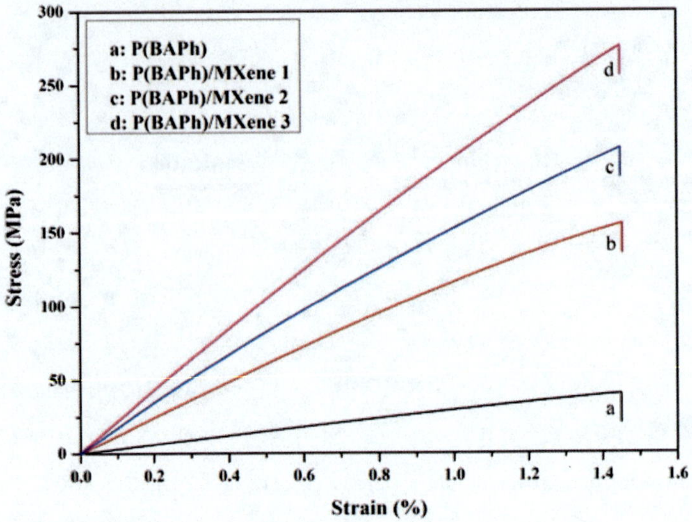

Fig. 9.12 Tensile stress–strain curves of the P(BAPh) matrix and the P(BAPh)/MXene nanocomposites with 1, 2, and 3 wt.% nanofiller (Derradji et al. 2019)

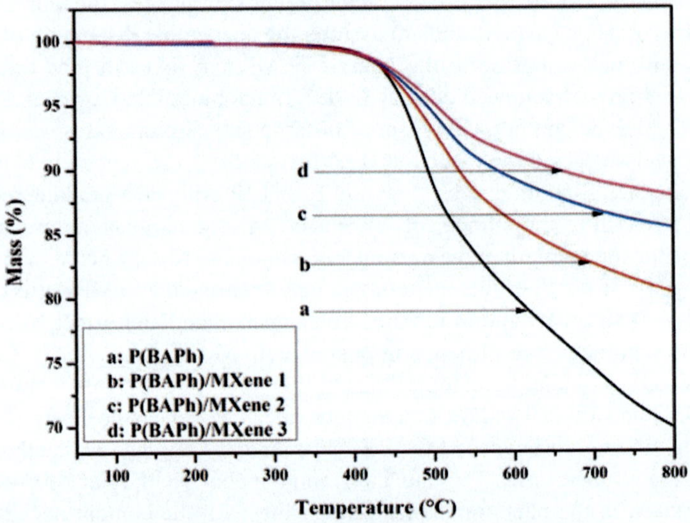

Fig. 9.13 TGA: thermal stability of the P(BAPh) matrix and the P(BAPh)/MXene nanocomposites with 1, 2, and 3 wt.% nanofiller (Derradji et al. 2019)

Table 9.1 Thermal properties of the P(BAPh)/MXene nanocomposites in nitrogen medium (Derradji et al. 2019)

Specimen	$T_{5\%}$ (°C)	$T_{10\%}$ (°C)	Y_c (%, 800 °C)
P(BAPh) matrix	463	493	69.8
P(BAPh)/MXene (1%)	477	530	80.3
P(BAPh)/MXene (2%)	491	576	85.4
P(BAPh)/MXene (3%)	502	649	87.8

$T_{5\%}$: 5% weight loss temperature. $T_{10\%}$: 10% weight loss temperature

in an inert atmosphere reaches 649 °C, which is by 156° higher than the analogous temperature when heated neat matrix.

9.5 Phthalonitrile/Tungsten Nanocomposites

Being designated for extremely exigent applications, the phthalonitrile resins need to be investigated for their behavior and shielding performances under high-energy ionizing radiations such as gamma rays. Since penetrating gamma rays can have a large damaging effect on instruments and humans, it is necessary to estimate the maximum allowable radiation doses as well as the screening ratios. Before, there have been no reports on the effect of any type of high ionizing radiation on the radiation damage and shielding behavior of the phthalonitrile materials. Derradji et al. (2018b, 2020) performed such a study for the first time. In addition, in Derradji et al. 2018b were investigated composites based on non-toxic tungsten nanoparticles of ca. 100 nm in size, modified by a silane coupling agent (aminopropyl trimethoxy silane) and introduced into the matrix at different amounts ranging from 30 wt.% to 50 wt.%. A bisphenol-A-based P(BAPh) resin was used as the typical matrix. The radiation doses of 50, 100, 150, and 200 kGy were applied using Co-60 gamma rays. The impact of the maximum absorption dose on the thermal and mechanical properties of the samples was determined. The screening ratios of the neat matrix and its tungsten-based nanocomposites were also investigated at different nanofiller loadings and sample thicknesses.

The samples were cured by a hot compression molding technique at temperatures of 240–340 °C. The cured nanocomposites were noted as P(BAPh)/W. The gamma radiation doses varied from 50 to 200 kGy. The initial and irradiated samples were characterized by FTIR, DMA, TGA, and in tension tests.

The gamma rays shielding efficiency of the P(BAPh)/W nanocomposites was studied by measuring the radiation dose absorbed through several sample thicknesses. Generally, the gamma rays interact with the materials either by absorption or scattering away. This interaction can be expressed as follows:

$$\frac{I}{I_0} = e^{-\mu x} \tag{9.1}$$

or

$$-Ln\left(\frac{I}{I_0}\right) = \mu x \tag{9.2}$$

where I_0 (herein equal to 1.17 meV) represents the radiation dose without the shielding material, I refers to the radiation dose through a thickness x(cm) of shielding material, μ is the total linear attenuation coefficient of a specific material for gamma rays in a specific energy. The screening or shielding ratio (S) is defined as follows,

$$S = \left(I_0 - \frac{I}{I_0}\right) \times 100 \tag{9.3}$$

A half-value layer (HVL) is used to a thickness where the dose after absorbing is half of the original.

$$\text{HVL} = Ln2/\mu \tag{9.4}$$

The primary goal of this research was to define the basic behavior of the phthalonitrile resin under highly ionizing gamma rays. The impact of different doses of gamma radiation on the storage modulus and T_g of the neat phthalonitrile matrix was estimated. It was shown that the storage modulus at 50 °C was 1.8 GPa and $T_g = 380$ °C. After 150 kGy exposure, these two parameters increased up to 2.1 GPa and 412 °C, respectively. Further increasing the latter to 200 kGy resulted in a slight decrease in the thermomechanical performances.

FTIR spectra allowed to interpret this dual effect. On the one hand, it turned out that the curing regime used did not result in sufficient completion of this process: the low-intensity absorption band of nitrile groups 2234 cm^{-1} remained in the spectrum of the initial material. It, however, practically disappeared after the radiation treatment with a dose of 150 kGy. This means that radiation activated and completed the polymerization process, i.e., played to a certain extent the role of post-polymerization at higher temperatures. Indeed, as follows from the data in Table 7.1, the same polymer matrix had $T_g = 389$ °C in the cured state and $T_g = 446$ °C after the post-curing procedure at high temperature.

Thus, it was confirmed that the gamma radiation doses of no more than 150 kGy effectively enhanced the crosslinking degree of the matrix and, as a result, the thermomechanical properties were improved. On the other hand, however, after exposure to a radiation dose of 200 kGy, the intensity of the 1385 cm^{-1} absorption band increased, indicating the beginning of destructive processes in the polymer network; this manifested in the decrease of the storage modulus and T_g values.

TGA showed that the thermal stability of the P(BAPh) resin was virtually unchanged after gamma irradiation with a dose of 200 kGy. The tensile stress–strain curves of the P(BAPh) resin irradiated at various radiation doses of 50, 100, 150, and 200 kGy showed that the strength was somewhat improved, from 54 to 66 MPa, at the doses of 150–200 kGy since the strength is closely related the crosslinking degree of the network. The strain at failure did not change after irradiation.

The following Figs. 9.14, 9.15, and 9.16 illustrate the important role of introducing tungsten nanoparticles into this matrix for the gamma rays shielding efficiency of the highly filled P(BAPh)/W nanocomposites. Figure 9.14 shows the parameter – Ln(I/I_0) as a function of the sample thickness for the neat P(BAPh) matrix and its highly filled tungsten-reinforced nanocomposites with 30, 40, and 50 wt.% nanofillers. One can see a particularly beneficial effect of high nanotungsten content on the efficiency of protective properties of the nanocomposite against gamma radiation. Really, the indicated parameter is equal for the neat matrix 0.2 at the thickness of a protective layer of 2 cm and 0.6 at its thickness of 8 cm, but for the nanocomposite with a tungsten content of 50 wt.% this parameter increases to 0.5 at the thickness of 2 cm and to 2.2 at the thickness of 8 cm.

Of interest is Fig. 9.15 where the screening or shielding ratio S and the total linear attenuation of a specific material for gamma rays μ are presented as a function of the nanotungsten content in the composites with the 2 cm thickness. Thus, the shielding ratio S increased from 17% for the neat matrix to 42% for the nanocomposite with 50 wt.% tungsten nanoparticles.

At last, a half-value layer (HVL) as a thickness where the dose after absorption is half of the original one is shown in Fig. 9.16 as a function of the nanofillers' loading. One can see that this parameter decreases from 10 cm in the case of the neat matrix down to 2 cm for the nanocomposite with 50 wt.% of tungsten nanoparticles.

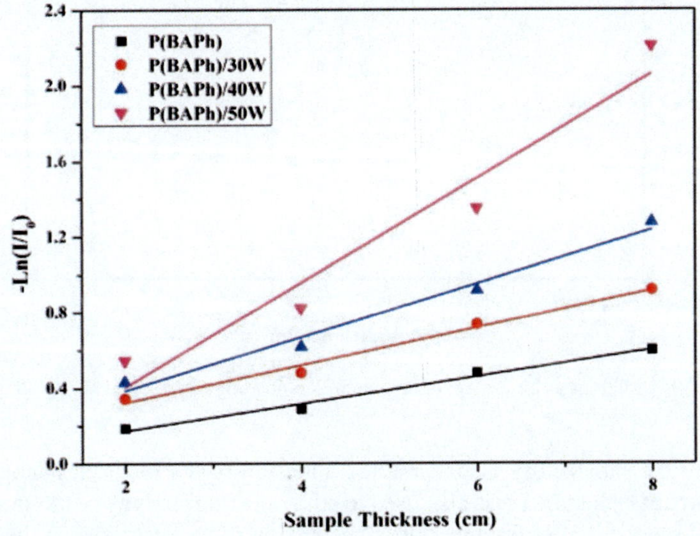

Fig. 9.14 Ln(I/I_0) as a function of the sample thickness for the neat P(BAPh) resin and its tungsten-reinforced nanocomposites (Derradji et al. 2018b)

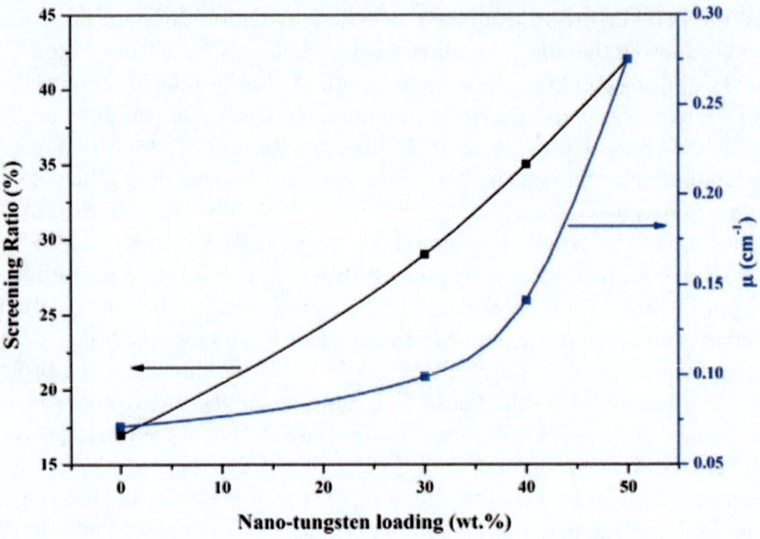

Fig. 9.15 S and μ variations as a function of the tungsten nanofillers loading (Derradji et al. 2018b)

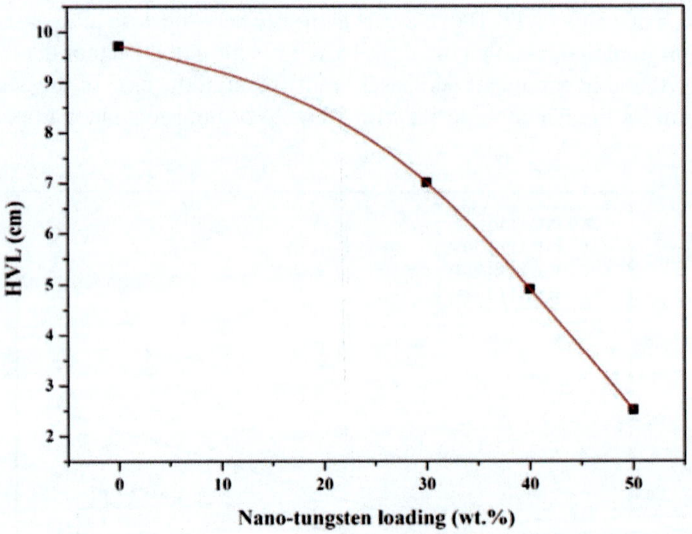

Fig. 9.16 HVL variations as a function of the nanofillers loading (Derradji et al. 2018b)

To prove the satisfactory dispersion and adhesion of the tungsten nanoparticles within the matrix, the SEM was also used to study the morphology of the samples in their initial state and after fracture. Good dispersion of nanoparticles in the matrix, satisfactory adhesion of the matrix to nanoparticles modified with silane coupling agent (no debonding), and a sharp change in the character of mechanical fracture in nanocomposites were confirmed.

It is also important to point out that the SEM was performed before and after the radiation test; however, no morphology change due to the radiation action was depicted. These observations are in good agreement with the findings especially since the P(BAPh) matrix can withstand high gamma rays radiation doses.

Thus, the study (Derradji et al. 2018b) showed that the P(BAPh) matrix and especially its highly filled tungsten-containing nanocomposites were nearly not affected by gamma rays radiation dose as high as 200 kGy.

That confirms that the remarkable performances of the phthalonitrile matrix and its nanocomposites are not limited to the thermal and mechanical properties only but can be extended to the gamma rays radiation and shielding resistances, especially in the case of tungsten-containing nanocomposites.

9.6 Phthalonitrile/Graphite Nanoplatelets Nanocomposites

Lei et al. (2012) have prepared and studied the exfoliated graphite nanoplatelets (xGnP) filled 4,4′-Bis (3,4-dicyanophenoxy) biphenyl (BPh) nanocomposites.

As it is well known, graphite is a natural layered material consisting of one-atom-thick sheets of carbon (graphene). The carbon atoms are bonded covalently in a hexagonal arrangement within the layer. These layers are bound to each other by weak van der Waals forces. Along the basal plane, graphite could exhibit high modulus, excellent electrical and thermal conductivities as well a low coefficient of thermal expansion.

Expandable graphite can be expanded up to hundreds of times more than its initial volume at high temperatures, resulting in the separation of the graphene sheets at the nanoscopic level along the c-axis of the graphene layers. Recently, a new nanoreinforcement, exfoliated graphite nanoplatelets (xGnP), has been under investigation (Kalaitzidou et al. 2007; Jiang and Drzal 2010).

In (Lei et al. 2012) xGnP was prepared by intercalation of natural graphite followed by rapid exfoliation in a microwave environment. Expanded graphite (EG) consisted of graphite nanosheets with a thickness of less than 100 nm thick. Then, they were mechanically grounded to form individual xGnP. The diameter of most xGnP varied from 5 to 20 mm, whereas the thickness of xGnP was in the range of 10–30 nm. It means that their aspect (diameter to thickness) ratios were as high as 10^2–10^3.

The BPh/xGnP prepolymer systems were cured at 250–375 °C. As a result, the BPh/xGnP nanocomposites with different xGnP contents, 0, 2, 5, 10, and 15 wt.%, were obtained. SEM, TGA, and mechanical and electrical resistivity measurements have been performed.

Figure 9.17 shows the mechanical properties of the nanocomposites with various xGnP contents. The flexural strength and modulus of the nanocomposites increased with increasing xGnP content.

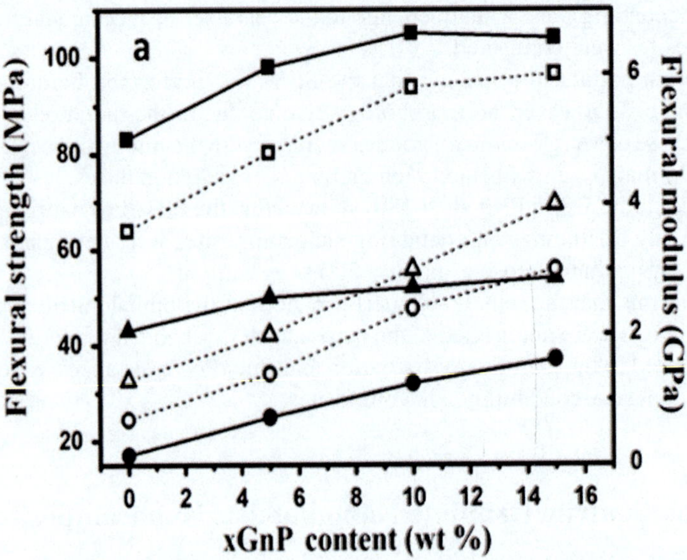

Fig. 9.17 Comparison of the flexural strength (closed symbols) and modulus (open symbols) between the BPh/xGnP (squares, Ref. (Lei et al. 2012)), and PP/xGnP-15 (triangles, Ref. (Kalaitzidou et al. 2007)) and HDPE/xGnP-15 (cycles, Ref. (Jiang and Drzal 2010))

The maximum values of mechanical properties were observed in the 10 wt.% xGnP-filled BPh systems, where about 27 and 69% increments of flexural strength and modulus were obtained in comparison with those of neat BPh matrix. The significant improvement in mechanical properties of the BPh matrix by xGnP was due to the excellent mechanical properties of xGnP as well as their large aspect ratio and good dispersion confirmed by SEM. This figure also contains, for comparison, similar results obtained for the xGnP-filled nanocomposites with polyethylene and polypropylene matrices. TGA showed also higher values of $T_{5\%}$ and $T_{10\%}$ for the nanocomposites than for the neat BPh matrix. Thus, $T_{10\%}$ could increase from 559° to 615 °C (Lei et al. 2012).

Of importance are the results of the electrical resistivity estimation of the nanocomposites in question because, unlike many other types of nanofillers, carbon-based nanofillers such as xGnP have the potential to produce conductive nanocomposites; this may be of interest in some applications. It is clear that the electrical conductivity of the nanocomposite can be achieved only with a certain "connectivity" of the carbon-based filler particles. The phthalonitrile matrix itself is an excellent insulating material, with electrical resistivity of almost 10^{14} Ω cm, whereas xGnP has the electrical characteristics similar to those for metallic or semimetallic materials. It should be noted that the percolation threshold for the electrical resistivity, i.e., the transition from an electrical insulator to an electrical conductor or semiconductor, depends very much on the geometry of conducting fillers.

It is expected that nanoparticles in the form of nanosheets can provide a lower percolation threshold value than spherical or elliptical nanoparticles when forming a "conductive network" in the polymer matrix. The higher the aspect ratio of the filler nanosheet, the lower the percolation threshold is expected. Recall that an average aspect ratio of xGnP in the work in question was as large as 500.

Figure 9.18 shows the electrical resistivity of the BPh/xGnP nanocomposites as a function of the xGnP content. The incorporation of the xGnP greatly decreases the resistivity of the BPh, from ~10^{14} Ω cm down to the value between 10^7 and 10^8 Ω cm, that is, to a value already inherent in semiconductors. The percolation threshold of the nanocomposite is located between 5 wt.% and 10 wt.% of xGnP. The electrical resistance decreases insignificantly with further increase of the xGnP content in the nanocomposite above 10 wt.%. A further increase in the xGnP content in the nanocomposite does not have a significant additional influence on the resistivity of the nanocomposite because the conductive network has already been formed.

Thus, it must be concluded that introduction of 5–10 wt.% xGnP into phthalonitrile matrix leads to its transformation from an electrical insulator into a material with semiconductor electrical resistivity.

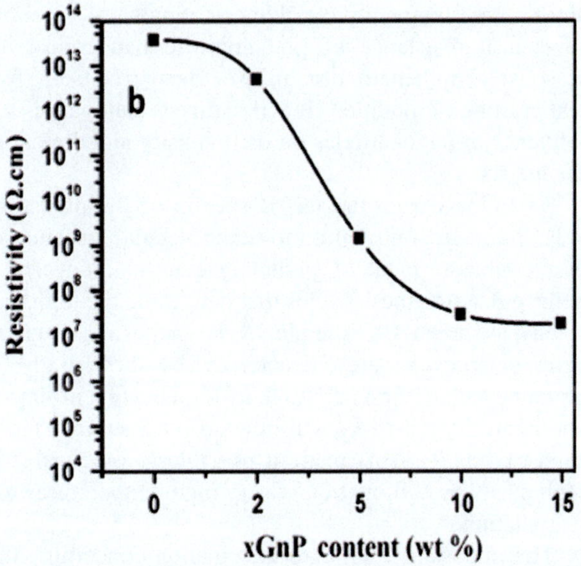

Fig. 9.18 Effect of the xGnP content on the electrical resistivity of the BPh/xGnP nanocomposites (Lei et al. 2012)

9.7 About the Origin of Super-Heat Resistance of Phthalonitrile Nanocomposites

Thus, from the foregoing in Chaps. 6–9 it is evident that phthalonitrile nanocomposites are characterized by the extraordinary resistance to the impact of high and ultra-high temperatures.

First, as indicated above, in the case of phthalonitrile nanocomposites containing functionalized MMT nanolayers or POSS nanoparticles, it was possible to achieve glass transition temperatures of T_g = 560–570 °C after their post-curing. Good thermal stability of the neat phthalonitrile matrix under a nitrogen atmosphere was confirmed by the $T_{5\%}$ and $T_{10\%}$ values of 464° and 493 °C, and a char yield after heating to 800 °C $Y_c = 69.8\%$. At BN nanofillers loading, the $T_{5\%}$ for the nanocomposite increased to the value of 551 °C and no $T_{10\%}$ was observed since the char yield at 800 °C $Y_c = 92.2\%$. The introduced MXene nanosheets effectively improved the thermal resistance of the P(BAPh) matrix: when heated in an inert atmosphere, $T_{10\%}$ value reached 649 °C, which was by 156° higher than the analogous temperature for the neat matrix.

Thus, in fact, the assumption that phthalonitrile nanocomposites are leading among high-temperature polymer composites is confirmed. It is obvious that the super-heat resistance of phthalonitrile nanocomposites is determined by three reasons: (a) high thermal stability of the matrix itself; (b) good dispersivity and chemical grafting of modified (functionalized) nanoparticles to the matrix, and (c) strong influence of nanoparticles on the intensity and direction of destructive processes in the matrix.

As follows from the far-IR spectra of the phthalonitrile matrix (Figs. 6.4 and 7.2), the main contribution to the molecular structure of this heterocyclic network is undoubtedly made by phthalocyanine macrocycles. There are strong reasons to believe that it is these cycles that determine the high thermal stability of the matrix. Really, as shown, for example, by Wagner et al. (Wagner et al. 1982), phthalocyanine macroheterocycles are characterized by surprisingly high thermal stability. Thus, phthalocyanine is resistant to short-term heating in air to temperatures of 400–500 °C, and in an oxygen-free environment (in a vacuum or in an inert gas environment)—even to 900 °C. This made it possible to obtain this heterocyclic substance with high purity by sublimation at very high temperatures without its melting or thermal degradation.

The important additional information concerning the high thermal stability and thermal degradation processes in phthalonitrile nanocomposites was obtained by DSC, FTIR, XRD, and X-ray photoelectron spectroscopy (XPS) by Li et al. (2018), as well as by IR spectroscopy by Yakushev et al. (2103).

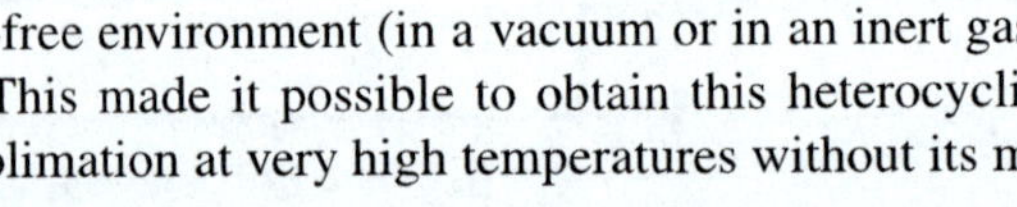

Li et al. (2018) prepared and studied the phthalonitrile nanocomposite based on BPh monomer (see a formula) with adding 0.5 wt.% epoxy cyclohexyl POSS derivative (EP0408, Hybrid Plastics, USA). The neat prepolymer mixture and POSS-containing prepolymer mixture were cured in successive heating treatments of 260° (8 h), 300° (8 h), and 325 °C (8 h) under an inert atmosphere of nitrogen. Then, the samples were post-cured at 350–375 °C under the same atmosphere. Below the prepared samples are called as neat polymer and POSS-containing polymer. The samples were tested in initial unsintered state or after sintering for 3 h in either a nitrogen or air medium at 500°, 600°, 700°, or 800 °C.

Figure 9.19 shows IR spectra of the neat phthalonitrile matrix (a) and POSS-containing nanocomposite (b) in the initial unsintered state and after sintering at temperatures of 500°, 600°, or 800 °C in inert atmosphere. A comparison of these spectra allows one to draw several conclusions.

First, with increasing temperature of sintering in both cases a decrease in intensity and even disappearance of most absorption bands, a certain "depletion" of the IR spectra is observed. These changes are natural, since they are caused by the processes of thermal degradation that have started.

Secondly, the absorption band of nitrile groups (–C≡N stretching vibrations) 2230 cm^{-1} is clearly present in the spectra of unsintered samples, although they have passed the post-curing procedure. This indicates an incomplete polymerization process and is consistent with our experimental data given above in Table 6.1. This band disappears after sintering.

Further, as a result of sintering, the low-intensity absorption band of isoindoline cycle vibrations at 1720 cm^{-1} disappears and the intensity of the triazine cycle vibrations band from 1560 cm^{-1} to 1510 cm^{-1} dramatically decreases.

At the same time, a contrasting behavior (different for the neat matrix and the nanocomposite) can be seen for the absorption band of phthalocyanine cycles at ca. 3450 cm^{-1} (NH stretching vibrations in the phthalocyanine). Figure 9.19 shows that in the case of the neat polymer matrix, the intensity of this band increases markedly after sintering at 500 °C, but then decreases significantly after heating at 600° and 800 °C. A different picture is observed after the same thermal treatments of POSS-containing nanocomposite. In this case, the intensity of the phthalocyanine absorption band does not decrease at all regimes of heat treatment; moreover, after sintering at 600 °C this band increases. This result is consistent with the effects of increasing thermal stability of the material due to embedded nanoparticles. It also indicates the appearance of new phthalocyanine cycles at the expense of annihilated triazine and isoindoline cycles.

Yakushev et al. (2021) confirmed the determining role of phthalocyanine macro-heterocycles, the main element of the molecular structure of the phthalonitrile matrix network, and the positive influence of the covalent incorporation of modified nanoparticles (5 wt.% amino-MMT nanolayers) into it to achieve ultra-high thermal stability effect.

Figure 9.20 presents the FTIR spectra of the cured BAPhN/amino-MMT (5 wt.%) nanocomposite in its initial state and after heating with the rate of 20 °C/min up to 900 °C in nitrogen atmosphere (TGA experiment). There are several noteworthy

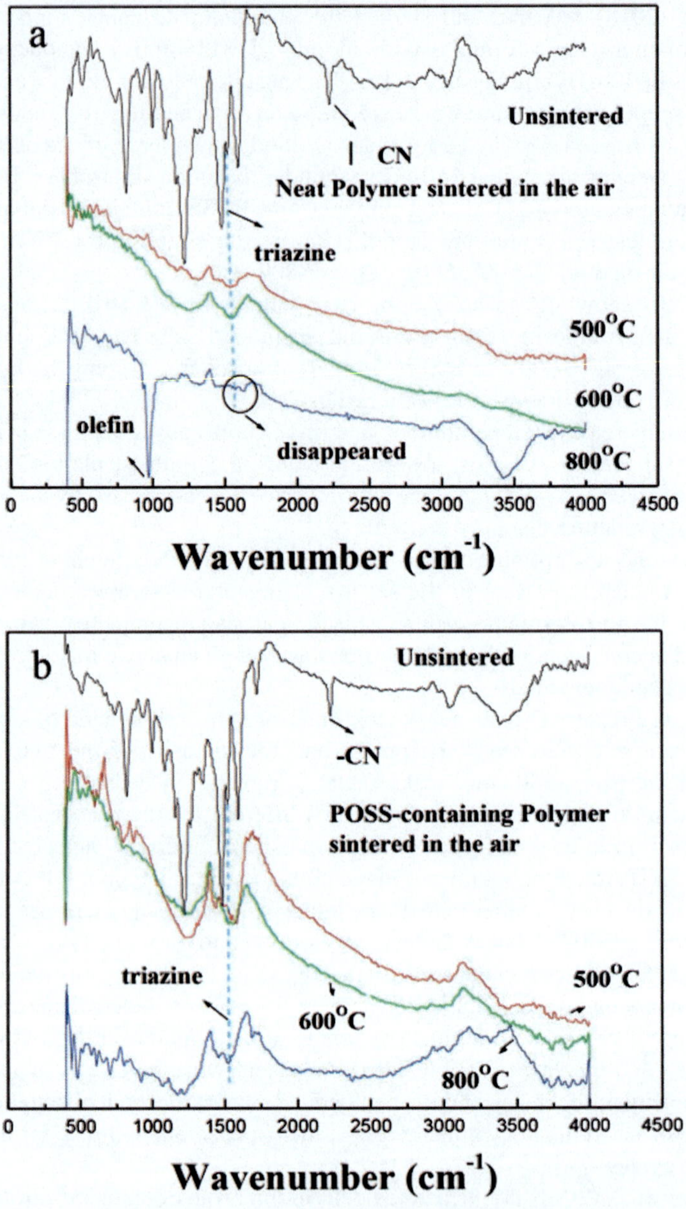

Fig. 9.19 Comparison of IR spectra of the neat polymer (**a**) and POSS-containing polymer (**b**) when sintered in N_2 atmosphere at different temperatures for 3 h, and of the unsintered control samples (Li et al. 2018)

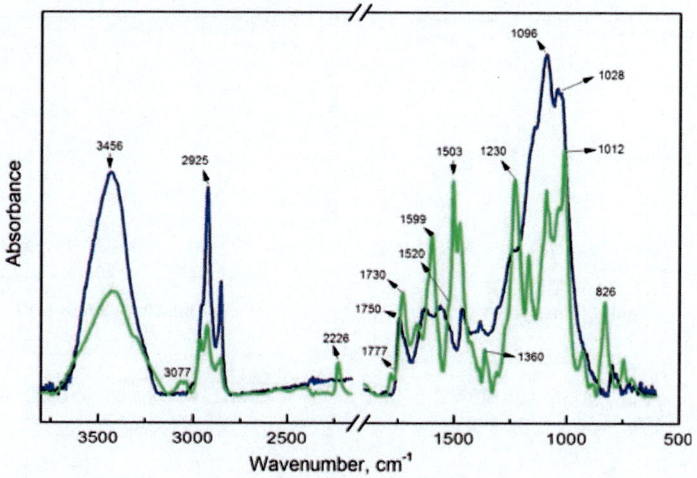

Fig. 9.20 FTIR spectra of the cured BAPhN/amino-MMT (5 wt.%) nanocomposite before (green line) and after heating at 900 °C in nitrogen medium (blue line) (Yakushev et al. 2021)

changes in the spectrum as a result of such high-temperature exposure. First of all, it was a remarkable fact that the nanocomposite retained its integrity after this treatment. Further, the intensities of absorption bands at 1360 cm^{-1} and 1520 cm^{-1} (triazine cycles) decreased, and the band at 1503 cm^{-1} (isoindoline cycles) disappeared at all which indicated the intense thermal degradation of these cycles. At the same time, the intensities of the absorption bands at 1010 cm^{-1} (displaced to 1028 cm^{-1}) and 3456 cm^{-1}, characterizing phthalocyanine cycles, increased significantly. This result led to conclusion that at the super-high-temperature pyrolysis of phthalonitrile nanocomposite, the destruction of the matrix triazine and isoindoline cycles is accompanied with their partial transformation into very stable phthalocyanine macrocycles.

Figure 9.21 shows the O 1 s XPS spectra of the neat polymer and POSS-containing polymer when sintered in air at 600 °C and 800 °C for 3 h and the unsintered samples. In this case, sintering led to the passing (to a greater or lesser degree) of the thermo-oxidative degradation of materials. These spectra characterized different electronic states of O element. One can see that the given XPS spectra differ significantly depending on both the material composition (in the absence or presence of nanoparticles) and the thermal prehistory. Moreover, they are also different for these two materials in the unsintered state, i.e., after post-curing at 375 °C. In other words, the introduction of 0.5 wt.% POSS nanoparticles exerts a great influence on the nature, direction, and degree of thermos-oxidative destructive processes in the nanocomposite.

For the unsintered samples, neat polymer and POSS-containing polymer, two peaks were observed at 533.1 eV and (531.2–531.6) eV which corresponded to Ph-O-Ph and -C = O bonds, respectively (Fig. 9.21a, b). Meanwhile, the contribution of

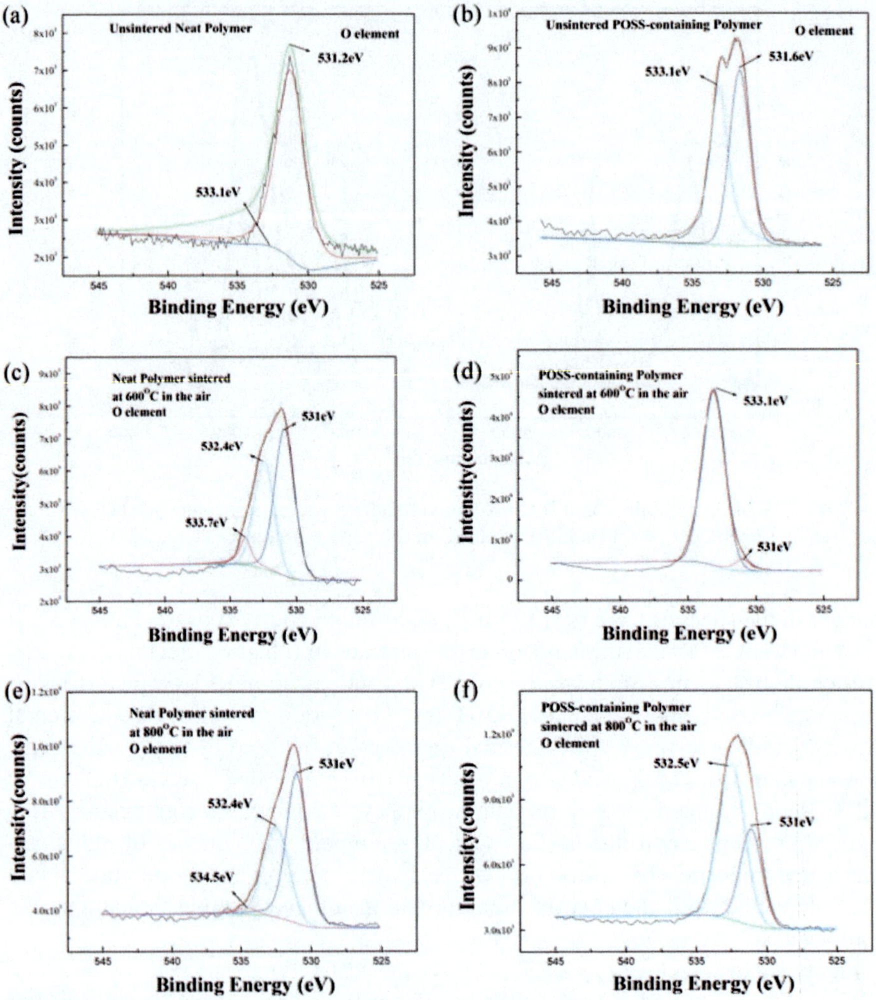

Fig. 9.21 XPS spectra of O 1 s of the neat polymer and POSS-containing polymer when sintered in air at 600 °C and 800 °C for 3 h and the unsintered control samples. **a** Unsintered neat polymer. **b** Unsintered POSS-containing polymer. **c** Neat polymer sintered at 600 °C. **d** POSS-containing polymer sintered at 600 °C. **e** Neat polymer sintered at 800 °C. **f** POSS-containing polymer sintered at 800 °C (Li et al. 2018)

the first of these bonds (peaks in the spectrum) inherent in the initial BPh monomer (see above) was significantly greater for the nanocomposite.

A large difference in the O 1 s XPS spectra of the neat matrix and nanocomposite was observed after their sintering at 600 °C (Fig. 9.21c, d). In this case, three peaks were observed at 531 eV, 532.4 eV, and 533.7 eV, corresponding to –OH, –C = O, and C-O bonds, respectively, for the neat matrix. At the same time, after this thermal treatment of the nanocomposite, a very intense peak at 533.1 eV corresponding to

Ph-O-Ph bonds strongly prevailed in its O 1 s XPS spectrum. This clearly indicates that the nanocomposite is much more resistant to thermo-oxidative degradation under these conditions compared to the neat matrix.

At last, during sintering at 800 °C, the destructive thermo-oxidative processes prevailed both in the neat matrix and in the nanocomposite, although a difference in their character is also noted in these conditions. In fact, two sharply pronounced peaks in the O 1 s XPS spectra, at 531 eV and 532.4–532.5 eV, corresponding to the –OH and –C = O groups were observed; however, their intensities differed significantly in the case of neat polymer and POSS-containing polymer (Fig. 9.21e, f).

The data of Li et al. (2018) to estimate the activation energies of the thermal degradation process of the neat matrix and POSS-containing nanocomposite are given in Fig. 9.22. For this purpose, thermogravimetric analysis was performed when the samples were heated from ambient temperature to 1000 °C in the air atmosphere with different heating rates (10, 20, 30, and 40 °C/min). Based on these experimental data, the activation energies of the thermal degradation process (mass loss) were calculated as a function of the degree of material degradation.

Figure 9.22 presents these activation energies obtained for the neat phthalonitrile matrix and the POSS-containing nanocomposite. In general, the activation energy varied from 36 kJ/mol to 118 kJ/mol, and it is obvious that the thermal degradation of the nanocomposite is the more slow process occurring with overcoming the higher activation barriers than that of the neat matrix. At all degrees of degradation, the activation energies of this process for the nanocomposite are higher by 25–30 kJ/mol. Consequently, these data also confirm the retarding influence of embedded nanoparticles on the thermal degradation of the phthalonitrile matrix.

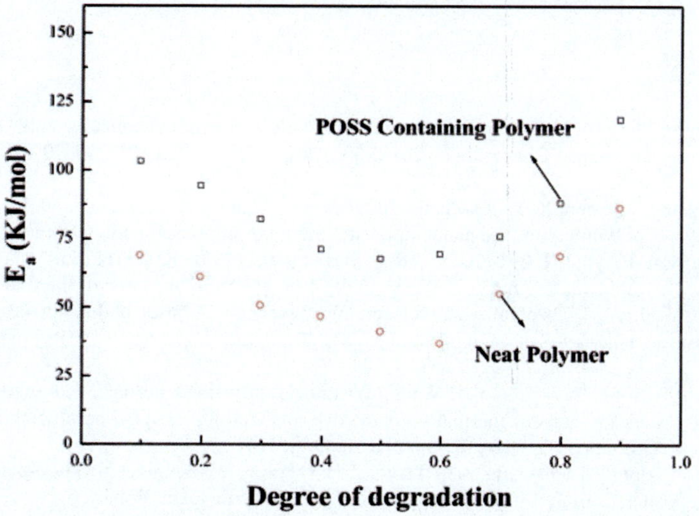

Fig. 9.22 Plots of activation energy as a function of the degree of degradation for the neat polymer and the POSS-containing polymer (Li et al. 2018)

Finally, of interest is a comparison of XRD patterns obtained for neat phthalonitrile matrix and for POSS-containing nanocomposite in the initial cured state and after sintering in air at 600°, 700°, and 800 °C (Li et al. 2018). The neat matrix in the initial state exhibits a completely amorphous structure and broad peak with the maximum at $2\theta = 23.0–23.8°$. At 600 °C, there was an actual increase in the intensity of this peak and a distinguishable manifestation of a peak with a maximum around $2\theta = 46°$, indicating some ordering of the structure. However, with further temperature increase the intensity of the main peak decreases, and at 800 °C actually, both peaks disappear, indicating that the structure is disordered, obviously due to the degradation process. A different picture was observed for the POSS-containing nanocomposite. In this case, the XRD patterns show more intense peaks, and they definitely do not disappear with increasing temperature. A shift to a higher diffraction angle 2θ, from 20.4° to 24°, with increasing temperature could be seen. This result is consistent with the proposition that the covalent embedding of POSS nanoparticles into the polymer matrix actually increased the resistance of the material to the disordering and destructive action of ultra-high temperatures.

In conclusion, it should be recognized once again: the whole complex of published experimental data presented in this book, including the supplementary data of this Sect. 9.7, definitively confirm, in our opinion, the leading position of phthalonitrile nanocomposites in the field of the super-heat resistant constructional and functional polymeric materials. The positive influence of covalently incorporated nanoparticles on thermal stability of materials is due to a certain suppressive effect of nanoparticles on matrix dynamics and their blocking effect by creating the steric obstacles to molecular movement and diffusion of decomposition products.

References

M.W. Barsoum, The $M_{N+1}AX_N$ phases: a new class of solids: thermodynamically stable nanolaminates. Prog. Solid State Chem. **28**(1), 201 (2000). https://doi.org/10.1016/S0079-6786(00)00006-6

C. Chen, J. Wang, X. Chen, X. Yu, Q. Zhang, Improvement of thermal conductivity and mechanical properties for polyphthalonitrile nanocomposites via incorporating functionalized h-BN fillers. High Perform. Polym. **31**, 294 (2019). https://doi.org/10.1177/0954008318765066

M. Derradji, N. Ramdani, T. Zhang, J. Wang, Z.-W. Lin, M. Yang et al., High thermal and thermomechanical properties obtained by reinforcing a bisphenol-A based phthalonitrile resin with silicon nitride nanoparticles. Mater. Lett. **149**, 81 (2015). https://doi.org/10.1016/j.matlet.2015.02.122

M. Derradji, X. Song, A.Q. Dayo, J. Wang, W. Liu, Highly filled boron nitride-phthalonitrile nanocomposites for exigent thermally conductive applications. Appl. Therm. Eng. **115**, 630 (2017). https://doi.org/10.1016/japplthermaleng201612141

M. Derradji, A. Henniche, J. Wang, A.Q. Dayo, J.-H. Ouyang, W.-B. Liu et al., High-performance nanocomposites from Ti_3SiC_2 MAX phase and phthalonitrile resin. Polym. Compos. **39**, 3705 (2018a). https://doi.org/10.1002/pc.24401

M. Derradji, A. Zegaoui, Y.L. Xua, A. Wang et al., Toward advanced gamma rays radiation resistance and shielding efficiency with phthalonitrile resins and composites. Nuclear Instruments Methods

Phys. Res. Sect. b: Beam Interact. Mater. Atoms **421**, 13 (2018b). https://doi.org/10.1016/j.nimb.2018.02.017

M. Derradji, D. Trache, A. Henniche, A. Zegaoui, A. Medjahed, A.F. Tarchoun, R. Belgacemi, On the preparation and properties investigations of highly performant MXene ($Ti_3C_2(OH)_2$) nanosheets-reinforced phthalonitrile nanocomposites. Adv. Compos. Lett. **28**, 1 (2019). https://doi.org/10.1177/2633366X19890621

M. Derradji, A. Zegaoui, A. Medjahed, W. Liu, A. Henniche, Hybrid phthalonitrile-based materials with advanced mechanical and nuclear shielding performances. Polym. Compos. **41**(1), 134 (2020). https://doi.org/10.1002/pc.25352

P. Eklund, M. Beckers, U. Jansson et al., The $M_{n+1}AX_n$ phases: materials science and thin-film processing. Thin Solid Films **518**(8), 1851 (2010). https://doi.org/10.1016/j.tsf.2009.07.184

A. Falin, Q. Cai, E.J.G. Santos et al., Mechanical properties of atomically thin boron nitride and the role of interlayer interactions. Nat. Commun. **8**, 15815 (2017). https://doi.org/10.1038/ncomms15815

A. Henniche, M. Derradji, J. Wang, W.-B. Liu, J.-H. Ouyang, A. Medjahed, High-performance polymeric nanocomposites from phthalonitrile resin and silane surface modified Ti_3AlC_2 MAX phase. High Perform. Polym. **30**(4), 427 (2018). https://doi.org/10.1177/0954008317699678

X. Jiang, L.T. Drzal, Multifunctional high density polyethylene nanocomposites produced by incorporation of exfoliated graphite nanoplatelets 1: morphology and mechanical properties. Polym. Compos. **31**, 1091 (2010). https://doi.org/10.1002/pc.20896

K. Kalaitzidou, H. Fukushima, L.T. Drzal, Mechanical properties and morphological characterization of exfoliated graphite–polypropylene nanocomposites. Compos. A **38**, 1675 (2007). https://doi.org/10.1016/j.compositesa.2007.02.003

Y. Lei, G.-H. Hu, R. Zhao, H. Guo, X. Zhao, X. Liu, Preparation process and properties of exfoliated graphite nanoplatelets filled bisphthalonitrile nanocomposites. J. Phys. Chem. of Solids **73**, 1335 (2012). https://doi.org/10.1016/j.jpcs.2012.07.003

X. Li, J. Wang, Y. Sun, D. Zhang, Z. Nan, J. Lei, The effect of thermal treatment on the decomposition of phthalonitrile polymer and phthalonitrile-polyhedral oligomeric silsesquioxane (POSS) copolymer. Polym. Degrad. Stab. **156**, 279 (2018). https://doi.org/10.1016/j.polymdegradstab.2018.09.012

D.V. Shtansky, K.L. Firestein, D.V. Golberg, Fabrication and application of BN nanoparticles, nanosheets, and their nanohybrids. Nanoscale **10**, 17477 (2018). https://doi.org/10.1039/C8NR05027A

N. Taloub, A. Henniche, L. Liu et al., Improving the mechanical properties, UV and hydrothermal aging resistance of PIPD fiber using MXene ($Ti_3C_2(OH)_2$) nanosheets. Compos Part b: Eng. **163**, 260 (2019). https://doi.org/10.1016/j.compositesb.2018.11.007

H.J. Wagner, R.O. Loutfy, C.K. Hsiao, Purification and characterization of phthalocyanines. J Mater Sci **17**, 2781 (1982). https://doi.org/10.1007/BF00644652

P.N. Yakushev, V.A. Bershtein, A.M. Fainleib, D.A. Kirilenko, O.G. Melnychuk, Ultra-heat resistant nanocomposites based on heterocyclic networks: structure, properties, the origin of thermal stability. J. Phys.: Conf. Ser. **2103**, 012113 (2021). https://doi.org/10.1088/1742-6596/2103/1/012113

C. Yu, J. Zhang, W. Tian, Y. Fan, Y. Yao, Polymer composites based on hexagonal boron nitride and their application in thermally conductive composites. RSC Adv. **8**, 21948 (2018). https://doi.org/10.1039/C8RA02685H

H. Zhang, L. Wang, Q. Chen et al., Preparation, the mechanical and anti-friction performance of MXene/polymer composites. Mater. Des. **92**(15), 682 (2016). https://doi.org/10.1016/j.matdes.2015.12.084

K. Zhang, Y. Feng, F. Wang, Z. Yang, J. Wang, Two-dimensional hexagonal boron nitride (2D-hBN): synthesis, properties, and applications. J. Mater. Chem. C **5**, 11992 (2017). https://doi.org/10.1039/C7TC04300G